Plasticity of Cold Worked Metals

A Deductive Approach

Plasticity of Cold Worked Metals
A Deductive Approach

A. Paglietti
University of Cagliari, Italy

WIT*PRESS* **Southampton, Boston**

Plasticity of Cold Worked Metals
A Deductive Approach

A. Paglietti
University of Cagliari, Italy

Published by

WIT Press
Ashurst Lodge, Ashurst, Southampton, SO40 7AA, UK
Tel: 44 (0) 238 029 3223; Fax: 44 (0) 238 029 2853
E-Mail: witpress@witpress.com
http://www.witpress.com

For USA, Canada and Mexico

WIT Press
25 Bridge Street, Billerica, MA 01821, USA
Tel: 978 667 5841; Fax: 978 667 7582
E-Mail: infousa@witpress.com
http://www.witpress.com

British Library Cataloguing-in-Publication Data

A Catalogue record for this book is available
from the British Library

ISBN-10: 1-84564-065-9
ISBN-13: 978-1-84564-065-1

LOC: 2006929310

Printed in Great Britain by Cambridge Printing

To Roberta,
whose love made this work possible

Contents

Preface

Plasticity Theory is a basic tool of structural analysis used to evaluate the ultimate strength and the post-elastic behaviour of ductile structures. Modern computing codes make use of it, often by exploiting sophisticated computational analysis and advanced kinematics for large deformation. Yet, all this valuable effort has its Achilles' heel: the evolution law of the yield surface, also referred to as the work-hardening rule.

The work-hardening rules most in use to model the post-elastic behaviour of elastic-plastic materials are, essentially, the kinematic rule, the isotropic rule or some combination – more or less linear – of these. None is capable of reproducing the dramatic changes that the elastic domain of real materials can exhibit when plastically strained. Other work-hardening rules are also available. They are invariably more complicated than the former and possibly even less adequate to represent the post-elastic behaviour of a material when it comes to general three-dimensional elastic-plastic deformation processes. Still, these processes are well within the reach of many commercial computer codes; which in principle should enable us to make any general elastic-plastic analysis we may care to do. The lack of realistic work-hardening rules, however, sets serious doubts on the validity of the results that can thus be obtained. As a consequence, we often have to confine our applications to particular classes of processes, reduce the range of admissible deformation and make frequent recourse to experimental validation.

The present book approaches the work-hardening problem in a new and deductive way. It starts from a few elementary, hardly questionable facts and works out their inescapable consequences, following them through until a practical solution is obtained. The analysis it presents will show that plastic yielding of ductile metals is essentially virgin yielding – no matter how or how strong the material was cold worked originally.

The book will also provide the reader with a definite tool to predict all possible subsequent yield surfaces of an elastic-plastic material, once its virgin surface is given. This means that the whole family of subsequent yield surfaces of a material is completely defined by its virgin yield condition. Such a result will be obtained in a purely deductive way, without introducing any special assumption about the way in which plastic deformation affects the evolution of the elastic range.

The emphasis of the book is on the class of materials that goes under the name of von Mises materials. Most metals belong to this class. It should not be difficult, however, to extend the main concepts of the book to other ductile materials, should the need arise.

Of course, it is one thing to determine the family of all possible subsequent yield surfaces that an elastic-plastic material can ever exhibit if properly cold worked, and quite a different thing to find out which of these surfaces should be associated to which cold working process. To do this, we need to know the evolution rule of the parameters upon which the surfaces of this family depend. An evolution law that is fairly general and fits to practical applications will be proposed in Ch. 7. The problem here is, however, a constitutive one. As a consequence, the realm of deductive reasoning must be abandoned, to give way to the many, different and sometimes contradictory experimental data obtained from different materials, or even from different experiments on the same material. As any experimentalist in this field knows only too well, the main difficulty here originates from the fact that the evolution of the yield surface is essentially strain history dependent. As such, any attempt at finding a reasonably simple evolution rule requires us to first of all extract the most representative variables from the jungle of all possible strain histories.

The theory developed in this book will enable us to spot these variables. The evolution law proposed in Ch. 7 will then try to strike a balance between the complexity caused by the strain history dependence, on the one hand, and the need for a practical tool to be handled by numerical codes on the other. The adequacy of the proposed law is evaluated in Ch. 9, where the predictions from the present theory are compared with the results taken from some well-known experimental works.

The present work grew gradually over a span of more than 10 years. On various occasions, I communicated parts of it at scientific meetings and conferences. As the material grew, however, it became evident to me that it worked like the cogwheels of a clock: once split apart, the ticking would come to a stop. In the end, I surrendered and settled for this monograph.

The book is addressed to research workers and advanced students in Mechanical Engineering, with a standard background in Continuum Mechanics and Plasticity Theory. These readers will find in it all the material they need for the applications. In particular, Appendix B is especially written to facilitate practical handling of rotation tensors. Such tensors find extensive use in the analysis of subsequent yielding, as the latter is closely linked to acquired anisotropy of the elastic limit.

I tried to organize the text so that it could also be of use to software engineers aiming at embodying the new theory in their computing codes. In doing so, they could use the work-hardening rules I present in this book, or exploit the general results concerning the family of all possible yield surfaces to propose better and perhaps simpler work-hardening rules.

The experimental scientist will benefit from this book as well. The general analysis it presents will serve him as a guide through the maze of experimental data. It will provide a new and most effective tool to give order, sense and perspective to the experimental findings and to direct new experimental research. The latter is still much needed in spite of so many years of valuable experimental activity, which unfortunately was partly marred by the lack of adequate understanding of the theoretical basis of subsequent yielding.

Aiming at an essential booklet, I confined the references to those I actually used in developing the theory it contains. They include the lucid and thoughtful treatise by Lubliner [1] and the excellent monograph by Paul [2]. In them the reader will find all the necessary background and much more. The reader is also referred to Lubliner's book, as well as to any recent textbook on Plasticity Theory, for a more complete coverage of the state of art of the discipline. It should be said, however, that the theory to be presented here is largely independent of previous works on the same topic.

Finally, but not least of all, I would like to express my thanks to the many friends and colleagues who encouraged me to undertake this work and gave me their suggestions and advice. My former student, Dr Giorgio Carta, was most helpful in reading the whole manuscript and providing many of the calculations needed to write Ch. 9. Dr Michele Brun kindly helped me to correct some formulae in the first part of the book. Mrs Claire Head checked the manuscript most efficiently giving valuable suggestions for a better English wording. The staff at Exeter Premedia Services did an excellent job of copy-editing, spotting a number of errors during the final typesetting. The WIT Press staff were always very friendly and cooperative in providing assistance during the entire process of preparation of this book. To all these people go my heartfelt thanks. Of course, any lapse or error the book may still contain is entirely my responsibility.

A. Paglietti
2007

1

Introduction

1.1 SOME BASIC PRELIMINARIES

To a greater or lesser extent, most metals exhibit ductile behaviour. They respond elastically if stress is kept within a certain region, variously referred to as elastic region, elastic range or elastic domain. Beyond that region, plastic deformation takes place. Usually, the deformation process modifies the elastic region itself and the material exhibits different subsequent elastic regions as the process proceeds. A stress space representation of the initial elastic range of a ductile material and some of its subsequent elastic ranges relevant to a generic deformation process is shown in Fig. 1.1.1. The figure should in fact be regarded as referring to a stress space of six dimensions, although for obvious reasons only two dimensions are represented here. Likewise, in the same figure the curves bounding the elastic regions are in fact hyper-surfaces in a six-dimensional space. They are usually called *yield surfaces* (or limit surfaces).

A similar representation can be drawn in the space of total strain rather than in stress space. In this case we speak of strain space representation. In this representation the yield surface is expressed as a function of total strain rather than stress. Strain space representation gives information about total strain, but leaves both elastic strain and plastic strain indeterminate. This contrasts with stress space representation, which gives information about stress and elastic strain, since the latter is related one-to-one to stress. The

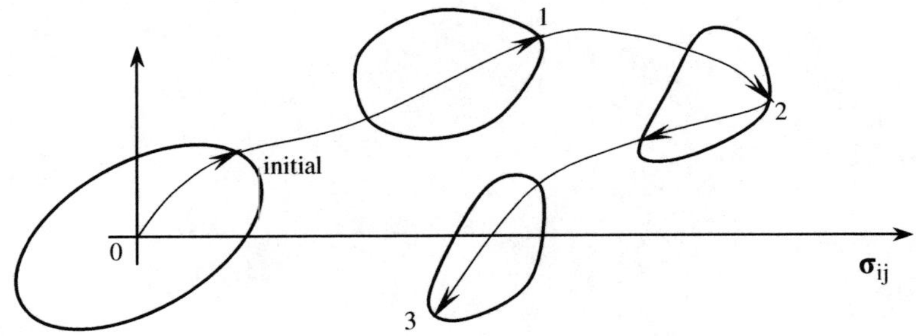

Fig. 1.1.1 Initial and subsequent elastic ranges in stress space.

latter property means that stress space representation is essentially the same thing as a representation in the space of elastic strain (or elastic-strain space representation). Neither of them, however, gives any information whatsoever about total strain or about its plastic component.

If the plastic component of strain is given, all the above representations are equivalent to each other, as elastic, plastic and total strain are all determined by any two of them. In this case the choice of the best representation to adopt is largely a matter of taste. In general, however, stress space and strain space representations convey different information about the process. The special instance considered in Fig. 1.1.2 may help to illustrate this point. Moreover, as will be discussed below, the stress space description of plasticity turns out to be unsuitable to deal with strain softening following plastic deformation.

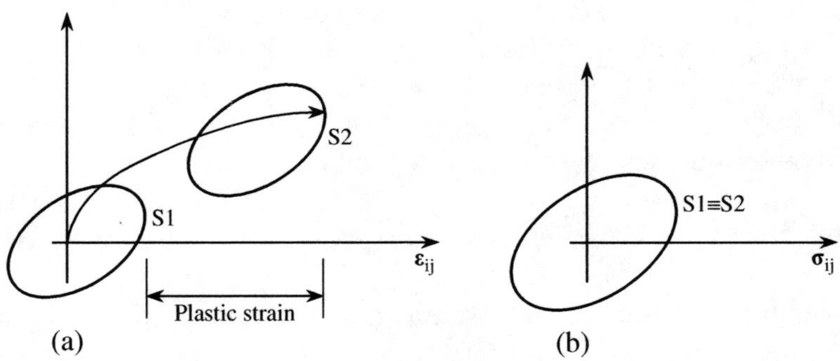

Fig. 1.1.2 Strain space and stress space representations of subsequent elastic regions of a material whose yield stress is unaffected by plastic deformation (*elastic-perfectly plastic material*): (a) two subsequent elastic regions in strain space; (b) stress space representation of the same regions.

The present book is concerned with macroscopic plasticity. The analysis it presents, however, is based on some considerations and hypotheses that are suggested by a few elementary facts concerning the physics of plastic deformation on a microscopic scale. These will be discussed in the next chapter. In the remaining sections of this chapter, we shall review the essential theoretical background of the macroscopic theory of plasticity.

1.2 GENERAL FRAMEWORK FOR CONSTITUTIVE EQUATIONS OF RATE-INDEPENDENT PLASTICITY

In isothermal conditions the state of an element of elastic-plastic material is defined by a certain number of independent variables (*state variables*) that can conveniently be grouped into two arrays, say $\boldsymbol{\xi}^{(e)}$ and $\boldsymbol{\xi}^{(p)}$. Array $\boldsymbol{\xi}^{(e)}$ contains the variables that help to describe the elastic state of the material. They will be different depending on whether we adopt the stress space or the strain space description of plasticity. If the stress space description is adopted, array $\boldsymbol{\xi}^{(e)}$ will include stress tensor $\boldsymbol{\sigma}$ or, equivalently, elastic strain tensor $\boldsymbol{\varepsilon}^e$. The latter is one-to-one related to $\boldsymbol{\sigma}$ through Hooke's law:

$$\boldsymbol{\sigma} = \mathbf{C}\boldsymbol{\varepsilon}^e \quad \text{or} \quad \sigma_{ij} = C_{ijlm}\varepsilon^e_{lm}, \tag{1.2.1}$$

where $\mathbf{C}$ is the elastic modulus tensor. On the other hand, if the strain space description is adopted, total strain tensor $\boldsymbol{\varepsilon}$, rather than $\boldsymbol{\sigma}$ or $\boldsymbol{\varepsilon}^e$, will appear among variables $\boldsymbol{\xi}^{(e)}$. In this case, however, the complete set of state variables must include plastic strain tensor $\boldsymbol{\varepsilon}^p$, otherwise the value of elastic strain and hence the value of stress cannot be determined. The validity of the widely accepted additive composition rule:

$$\boldsymbol{\varepsilon} = \boldsymbol{\varepsilon}^e + \boldsymbol{\varepsilon}^p \tag{1.2.2}$$

will be assumed throughout this book.

The second array of variables, namely $\boldsymbol{\xi}^{(p)}$, contains the inelastic variables. They may include the plastic strain tensor $\boldsymbol{\varepsilon}^p$, as well as other tensor and scalar variables related to plastic deformation and to plastic deformation history. The accumulated plastic work may appear in this array. All the variables that belong to $\boldsymbol{\xi}^{(p)}$ remain constant when the material behaves elastically. This occurs if its stress or elastic strain is kept in the elastic range, that is, within the yield surface. The latter is defined by a relation of the kind:

$$f(\boldsymbol{\xi}^{(e)}, \boldsymbol{\xi}^{(p)}) = 0, \tag{1.2.3}$$

where $\boldsymbol{\xi}^{(e)}$ stands for $\boldsymbol{\sigma}$ or $\boldsymbol{\varepsilon}$ depending on whether we adopt the stress space or the strain space description.

Function $f(\boldsymbol{\xi}^{(e)}, \boldsymbol{\xi}^{(p)})$ appearing in eq. (1.2.3) is usually referred to as the *yield function*. Here, we adhere to the usual convention of taking the sign of this function in such a way that the elastic region is contained *within* the yield surface and, accordingly, defined by $f(\boldsymbol{\xi}^{(e)}, \boldsymbol{\xi}^{(p)}) < 0$. In general, the yield surface will change during plastic deformation due to its dependence on $\boldsymbol{\xi}^{(p)}$. This produces what is commonly referred to as hardening or softening of the material, depending on whether it results in an increase or a decrease of its elastic limit.

1.2.1 Yield condition

There are two conditions that must be met for plastic deformation to take place. The first condition is that the state of the material must be on the yield surface. This means that eq. (1.2.3) must hold true. The second condition is that the stress increment must point outward from the yield surface or, equivalently, that its angle with the gradient of the yield surface must be less than $\pi/2$. This condition is usually referred to as *loading condition*. Since grad $f = \partial f/\partial \boldsymbol{\xi}^{(e)}$, this condition can be expressed by the following relation:

$$\frac{\partial f}{\partial \boldsymbol{\xi}^{(e)}} \dot{\boldsymbol{\xi}}^{(e)} > 0. \tag{1.2.4}$$

In the stress space description this inequality reads $(\partial f/\partial \sigma_{ij})\dot{\sigma}_{ij} > 0$, while in the strain space description the same inequality becomes $(\partial f/\partial \varepsilon_{ij})\dot{\varepsilon}_{ij} > 0$. Of course, function $f(\boldsymbol{\xi}^{(e)}, \boldsymbol{\xi}^{(p)})$ appearing in these inequalities is different depending on whether we are following the stress space or the strain space description. The two functions, however, can easily be related to each other, since $\boldsymbol{\sigma}$ and $\boldsymbol{\varepsilon}$ are related by the equation:

$$\boldsymbol{\sigma} = \mathbf{C}(\boldsymbol{\varepsilon} - \boldsymbol{\varepsilon}^{p}), \tag{1.2.5}$$

as immediately follows from eqs (1.2.1) and (1.2.2). Thus, if $f_1(\boldsymbol{\sigma}, \boldsymbol{\xi}^{(p)})$ denotes the yield functions of a material in the stress space description and $f_2(\boldsymbol{\varepsilon}, \boldsymbol{\xi}^{(p)})$ the yield function of the same material in the strain space description, we have $f_1(\boldsymbol{\sigma}, \boldsymbol{\xi}^{(p)}) = f_1[(\mathbf{C}(\boldsymbol{\varepsilon} - \boldsymbol{\varepsilon}^{e}), \boldsymbol{\xi}^{(p)}] \equiv f_2(\boldsymbol{\varepsilon}, \boldsymbol{\xi}^{(p)})$.

1.2.2 Consistency equation

Since the material responds elastically while in the elastic region, and since it cannot reach any point that is outside the yield surface without deforming plastically, it follows that any plastic deformation must occur *on* the yield surface. This means that condition (1.2.3) must hold true during any plastic deformation process, which also means that the all important *consistency equation*:

$$\dot{f} = \frac{\partial f}{\partial \boldsymbol{\xi}^{(e)}}\dot{\boldsymbol{\xi}}^{(e)} + \frac{\partial f}{\partial \boldsymbol{\xi}^{(p)}}\dot{\boldsymbol{\xi}}^{(p)} = 0 \tag{1.2.6}$$

must apply during such a process. This equation provides a fundamental relation between the time rates of $\boldsymbol{\xi}^{(e)}$ and $\boldsymbol{\xi}^{(p)}$ during plastic deformation.

An important consequence of eq. (1.2.6) concerns the possibility of describing the softening behaviour of a material. We may start by observing that from eqs (1.2.4) and (1.2.6) it follows that during plastic deformation we must have:

$$D = -\frac{\partial f}{\partial \boldsymbol{\xi}^{(p)}}\dot{\boldsymbol{\xi}}^{(p)} > 0, \tag{1.2.7}$$

the equation at the left-hand side of this expression being a definition of D. Inequality (1.2.7) means that a change in the inelastic variables makes the yield surface move (at least locally) in the direction of the outwardly oriented normal vector to the yield surface at the point where yield occurs. In the stress space description, this will increase the yield limit in the direction of the applied stress, thus producing work-hardening. Clearly, a movement of the yield surface in the opposite direction would produce the opposite effect, that is, softening. This would not be possible, though, as it would mean $D < 0$, in contrast to relation (1.2.7). When the stress space approach to plasticity is adopted, therefore, consistency equation (1.2.6) rules out plastic softening as a possible behaviour for an elastic-plastic material.

In the strain space approach, however, the situation is different. Here the above condition $D > 0$ means that the displacement of the yield surface will be in the direction of the applied strain $\boldsymbol{\varepsilon}$. This in no way excludes that the process can make the plastic component of strain change in such a way that the elastic strain at yield – that is, the yield limit of the material – decreases. In other words, when the strain space approach is adopted, condition $D > 0$ can accommodate both hardening and softening. This confers more generality to the strain space approach over the stress space one. The two approaches are equivalent, however, if applied to work-hardening materials.

1.2.3 Work-hardening rule

As already observed, the presence of $\boldsymbol{\xi}^{(p)}$ among the independent variables of the yield function means that the elastic domain will change due to plastic deformation. This feature is referred to generically as work-hardening. Accordingly, the set of evolution equations for the variables $\boldsymbol{\xi}^{(p)}$ that appear in eq. (1.2.3) will be referred to as the *work-hardening rule*. In the so-called rate-independent plasticity, these evolution equations are postulated in the form:

$$\dot{\boldsymbol{\xi}}^{(p)} = \dot{\lambda}\mathbf{g}(\boldsymbol{\xi}^{(e)}, \boldsymbol{\xi}^{(p)}). \tag{1.2.8}$$

This equation is understood to stand for an array of equations expressing the time rates $\dot{\xi}^{(p)}, \dot{\xi}_i^{(p)}, \dot{\xi}_{ij}^{(p)}, \ldots$ of each scalar, vector or tensor variable belonging to the set $\boldsymbol{\xi}^{(p)}$ as appropriate functions $g(\boldsymbol{\xi}^{(e)}, \boldsymbol{\xi}^{(p)})$, $g_i(\boldsymbol{\xi}^{(e)}, \boldsymbol{\xi}^{(p)})$, $g_{ij}(\boldsymbol{\xi}^{(e)}, \boldsymbol{\xi}^{(p)}), \ldots$, respectively. The scalar $\dot{\lambda}$ is supposed to be the same in all these equations. It represents the time derivative of an appropriate scalar function, which is not specified at the outset. The values of $\dot{\lambda}$ are determined by the requirement that the time rates of variables $\boldsymbol{\xi}^{(p)}$ should be such as to meet consistency equation (1.2.6). In order to obtain an explicit expression for $\dot{\lambda}$, we introduce eq. (1.2.8) into eq. (1.2.6), keeping into account yield conditions (1.2.3) and (1.2.4). In an abridged, and rather symbolic, notation the result can be expressed as:

$$\dot{\lambda} = \begin{cases} \dfrac{1}{-\dfrac{\partial f}{\partial \boldsymbol{\xi}^{(p)}}\mathbf{g}(\boldsymbol{\xi}^{(e)}, \boldsymbol{\xi}^{(p)})} \dfrac{\partial f}{\partial \boldsymbol{\xi}^{(e)}} \dot{\boldsymbol{\xi}}^{(e)} & \text{if } f = 0 \text{ and } \dfrac{\partial f}{\partial \boldsymbol{\xi}^{(e)}} \dot{\boldsymbol{\xi}}^{(e)} \geq 0; \\ 0 & \text{if } f < 0 \text{ or } f = 0 \text{ and } \dfrac{\partial f}{\partial \boldsymbol{\xi}^{(e)}} \dot{\boldsymbol{\xi}}^{(e)} < 0. \end{cases} \tag{1.2.9}$$

Equation (1.2.8) does not depend on the actual time rate at which the plastic deformation process occurs. This can immediately be verified by observing that a change of the time scale equally affects the time derivatives at both sides of this equation, thus effectively leaving it unaltered. Such a feature is the hallmark of rate-independent plasticity.

1.2.4 Plastic flow rule

The constitutive equations of an elastic-plastic material cannot be complete if they do not include the evolution law for plastic deformation, also referred to as *flow rule*. If $\boldsymbol{\varepsilon}^p$ is included among the variables $\boldsymbol{\xi}^{(p)}$, its evolution law will obviously be of the kind shown in eq. (1.2.8). That is:

$$\dot{\varepsilon}^{p}_{ij} = \dot{\lambda} h_{ij}(\boldsymbol{\xi}^{(e)}, \boldsymbol{\xi}^{(p)}), \tag{1.2.10}$$

where h_{ij} is an appropriate function of $\boldsymbol{\xi}^{(e)}$ and $\boldsymbol{\xi}^{(p)}$ and $\dot{\lambda}$ is given by eq. (1.2.9). On the other hand, if $\boldsymbol{\varepsilon}^{p}$ is not included in array $\boldsymbol{\xi}^{(p)}$, the assumption that the state of the material is completely defined by variables $\boldsymbol{\xi}^{(e)}$ and $\boldsymbol{\xi}^{(p)}$ means that such variables determine the value $\boldsymbol{\varepsilon}^{p}$. In this case, it is both reasonable and customary to assume that plastic strain does not depend explicitly on the elastic state variables. That is:

$$\varepsilon^{p}_{ij} = \varepsilon^{p}_{ij}(\boldsymbol{\xi}^{(p)}). \tag{1.2.11}$$

This obviously implies that:

$$\dot{\boldsymbol{\varepsilon}}^{p} = \frac{\partial \boldsymbol{\varepsilon}^{p}}{\partial \boldsymbol{\xi}^{(p)}} \dot{\boldsymbol{\xi}}^{(p)}. \tag{1.2.12}$$

From this equation and from eq. (1.2.8), it then follows that in this case too the evolution law of $\boldsymbol{\varepsilon}^{p}$ can be expressed in the form (1.2.10).

1.2.5 Plastic potential and normality rule

For special classes of materials, there may be a scalar function $p = p(\boldsymbol{\xi}^{(e)}, \boldsymbol{\xi}^{(p)})$ from which functions $h_{ij}(\boldsymbol{\xi}^{(e)}, \boldsymbol{\xi}^{(p)})$ appearing in eq. (1.2.10) can be obtained through a relation of the kind:

$$\mathbf{h} = \frac{\partial p(\boldsymbol{\xi}^{(e)}, \boldsymbol{\xi}^{(p)})}{\partial \boldsymbol{\xi}^{(e)}}. \tag{1.2.13}$$

Function p is then called *plastic potential.*

An important special instance occurs when this function coincides with the second principal invariant J_2 of the deviatoric part of $\boldsymbol{\sigma}$. That is:

$$p(\boldsymbol{\xi}^{(e)}, \boldsymbol{\xi}^{(p)}) = p(\boldsymbol{\sigma}) \equiv J_2(\boldsymbol{\sigma}), \tag{1.2.14}$$

where $J_2(\boldsymbol{\sigma})$ is the second invariant of the stress deviator. It is defined by:

$$J_2(\boldsymbol{\sigma}) = \frac{1}{2}[\mathrm{tr}(\mathbf{s}^2) - (\mathrm{tr}\ \mathbf{s})^2] = \frac{1}{2} s_{ij} s_{ij}, \tag{1.2.15}$$

s being the stress deviator:

$$\mathbf{s} = \boldsymbol{\sigma} - \bar{\sigma}\mathbf{I} \quad \text{or} \quad s_{ij} = \sigma_{ij} - \bar{\sigma}\delta_{ij}, \tag{1.2.16}$$

where scalar $\bar{\sigma}$ denotes mean stress:

$$\bar{\sigma} = \frac{1}{3}\mathrm{tr}(\boldsymbol{\sigma}) = \frac{1}{3}\sigma_{kk}. \tag{1.2.17}$$

Tensor $\bar{\sigma}\mathbf{I}$ appearing in eq. (1.2.16) is variously referred to as the *isotropic* or *spherical* or *hydrostatic* part of $\boldsymbol{\sigma}$. (Notice, however, that in the available literature it is not uncommon to find the second invariant defined as the opposite of that given by eq. (1.2.15). In this case, the present analysis still holds true provided that we change the sign of the far right-hand side of eq. (1.2.14).) A little calculation from eqs (1.2.15) and (1.2.16) shows that:

$$\frac{\partial J_2}{\partial \boldsymbol{\sigma}} = \mathbf{s} \quad \text{or} \quad \frac{\partial J_2}{\partial \sigma_{ij}} = s_{ij} \tag{1.2.18}$$

(for a formal treatment of the tensor derivatives of scalar invariants, the reader is referred to Leigh [3, Sect 4.7]). Therefore, when eqs (1.2.10) and (1.2.13) apply, from eqs (1.2.14) and (1.2.18), it follows that the flow rule (1.2.10) particularizes in the form (*Lévy-von Mises*):

$$\dot{\varepsilon}^p_{ij} = \dot{\lambda} s_{ij}, \tag{1.2.19}$$

showing that any plastic strain increment must be deviatoric. Plastic potential (1.2.14) implies, therefore, that plastic flow should occur at constant volume. This makes it particularly appealing in practice, because of the widespread experimental evidence that plastic deformation of many ductile metals is isochoric.

Quite often in the applications, plastic potential p is taken as coinciding with the yield function itself. That is $p(\boldsymbol{\xi}^{(e)}, \boldsymbol{\xi}^{(p)}) \equiv f(\boldsymbol{\xi}^{(e)}, \boldsymbol{\xi}^{(p)})$. In this case, the flow rule resulting from eqs (1.2.10) and (1.2.13) is referred to as *associated* to the yield function of the material. From eqs (1.2.10) and (1.2.13), the flow rule then becomes:

$$\dot{\boldsymbol{\varepsilon}}^p = \dot{\lambda}\frac{\partial f(\boldsymbol{\xi}^{(e)}, \boldsymbol{\xi}^{(p)})}{\partial \boldsymbol{\xi}^{(e)}}. \tag{1.2.20}$$

Since $\partial f/\partial \boldsymbol{\xi}^{(e)}$ is the gradient of the yield surface relevant to the given values of $\boldsymbol{\xi}^{(p)}$, this equation states that $\dot{\boldsymbol{\varepsilon}}^p$ must in this case be parallel to the outward unit normal to the yield surface. This is what is usually referred to as the *normality rule*. Such a rule is therefore a direct consequence of assuming a flow rule of the kind shown in eq. (1.2.19) associated with the yield function of the material.

1.2.6 Perfectly plastic materials

A material whose yield function does not depend on the inelastic variables is said to be an *elastic-perfectly plastic* material. To such a material, therefore, the condition:

$$\frac{\partial f}{\partial \boldsymbol{\xi}^{(p)}} = \mathbf{0} \tag{1.2.21}$$

applies. According to relation (1.2.7), this means that in this case D vanishes identically. Moreover, from the consistency condition (1.2.6), we can conclude that during the plastic deformation of a perfectly plastic material the condition:

$$\frac{\partial f}{\partial \boldsymbol{\xi}^{(e)}} = \mathbf{0} \tag{1.2.22}$$

must hold true too. By taking the limit of eq. (1.2.9) as $\partial f/\partial \boldsymbol{\xi}^{(p)}$ tends to zero, we can then infer that for perfectly plastic materials, $\dot{\lambda}$ is indeterminate when $f = 0$ and $(\partial f/\partial \boldsymbol{\xi}^{(e)})\dot{\boldsymbol{\xi}}^{(e)} \geq 0$. For these materials, therefore, the value of $\dot{\lambda}$ cannot be obtained from eq. (1.2.9). In this case, $\dot{\lambda}$ is usually determined by requiring that the deformation field of the body should be compatible with the body displacements at its boundary.

1.3 CONVEXITY OF THE YIELD SURFACE AND THE POSSIBLE LACK OF IT IN THE ELASTIC SUBRANGES

Let $\boldsymbol{\sigma}$ be any state of stress on the yield surface and let $\boldsymbol{\sigma}^*$ be any other state of stress within or on the same surface. Let, moreover, $\dot{\boldsymbol{\varepsilon}}^p$ be the plastic strain rate produced by an infinitesimal increment of stress starting from $\boldsymbol{\sigma}$. Inequality:

$$(\sigma_{ij} - \sigma^*_{ij})\dot{\varepsilon}^p_{ij} \geq 0 \tag{1.3.1}$$

is sometimes referred to as the *maximum dissipation postulate*. It has a long and distinguished history, being firstly proposed by von Mises (the reader may refer to Lubliner [1, Sect. 3.2.2] for more details). This inequality means that any stress variation, say $\Delta\boldsymbol{\sigma} = (\boldsymbol{\sigma} - \boldsymbol{\sigma}^*)$, that is capable of bringing the material to a point on the yield surface cannot be opposite to the strain rate $\dot{\boldsymbol{\varepsilon}}^{\mathrm{p}}$ the material will exhibit from that point, no matter how the stress is increased beyond the yield surface. (When making these kinds of geometric arguments, it is usual to consider stress and strain as six-dimensional vectors in a Euclidean space of six dimensions.)

Inequality (1.3.1) is also a part of the well-known Drucker's postulate, which provides a useful characterization of a class of *stable materials*, much in use in classical plasticity. A standard analysis of classical plasticity shows that inequality (1.3.1) implies that the yield surface should be convex and that, moreover, plastic strain rate $\dot{\boldsymbol{\varepsilon}}^{\mathrm{p}}$ should take place normal to this surface (*normality rule*). The latter property holds true with the proviso that the yield surface does not exhibit a singularity (or corner) at the point where yielding occurs.

Of course, materials for which the yield surface is not convex and/or the plastic flow rule – whether associated or not – does not obey the normality condition are perfectly admissible in nature. Such materials, however, do not meet inequality (1.3.1). For this reason, they may allow some form of material instability. For instance, they may flow plastically in mechanically insulated conditions (i.e. no exchange of work with the surroundings) at the expense of the elastic energy stored in the material itself.

Care should be exercised, however, before regarding the above results concerning convexity and normality as also applicable to the elastic subranges of the material. The latter are cross-sections of the full six-dimensional elastic range, where some of the independent elastic variables are either kept constant or constrained in a specific way. The boundaries of these subranges are in turn cross-sections of the yield surface of the material and are variously referred to as limit curves, limit surfaces, yield curves or even – when no confusion arises – yield surfaces themselves. Of course, no matter the dimensions of the space where the full yield surface is, any plane cross-section of it is convex too. The point to be noted here, however, is that quite often in practice, non-planar cross-sections of the full yield surface are in fact considered. In such a case, the resulting yield curve need not be convex even though the material is perfectly stable according to inequality (1.3.1).

A common instance of a non-planar cross-section of the full yield surface of the material is provided by the (σ, τ)-yield curves obtained from a specimen of material in a plane state of stress under tensile and shear forces. Usually, these curves are experimentally obtained from standard tension/torsion tests of thin-walled tubes. In a coordinate system (x_1, x_2, x_3) whose x_3-axis is normal to the plane of stress and the x_1-axis parallels the specimen axis, the stress tensor is given by:

$$\boldsymbol{\sigma} = \begin{bmatrix} \sigma & \tau & 0 \\ \tau & 0 & 0 \\ 0 & 0 & 0 \end{bmatrix}. \tag{1.3.2}$$

The two non-vanishing principal values of this tensor are given by:

$$\sigma_1, \sigma_2 = \frac{\sigma}{2} \pm \sqrt{\sigma^2/4 + \tau^2}, \tag{1.3.3}$$

and the direction of principal stress σ_1 makes an angle θ_3 with the x_1-axis given by:

$$\theta_3 = \frac{1}{2}\text{atan}\frac{2\tau}{\sigma}. \tag{1.3.4}$$

If the yield function of the material is not isotropic, as is likely after even a moderate plastic strain, the yield curves in the (σ, τ)-plane will change as the material element is rigidly rotated about the x_3-axis, while the applied stress is kept fixed. In the present case, the latter is completely defined by σ_1, σ_2 and θ_3 or, equivalently, by σ, τ and θ_3. From eq. (1.3.4) we see, however, that angle θ_3 is generally different for different values of σ and τ. This means that as σ and τ are changed, the principal directions of $\boldsymbol{\sigma}$ rotate by θ_3 about the x_3-axis. As far as the yield condition of the material is concerned, this is equivalent to keeping σ_1, σ_2 (or σ and τ) constant and rotating the element by angle $-\theta_3$ about the x_3-axis. Such a rotation of the material relative to stress, however, will in general modify its yield curve, if the yield surface is not isotropic. It must be concluded that for given values of the inelastic variables, the (σ, τ) – yield curves will in general depend on θ_3. They will accordingly be of the form:

$$t(\sigma, \tau, \theta_3) = t'(\sigma_1, \sigma_2, \theta_3) = 0, \tag{1.3.5}$$

depending on whether they are expressed as functions of σ and τ, or of σ_1 and σ_2. Such yield curves are not plane cross-sections of the yield surface of the material.

To justify the last claim, let us observe that for given inelastic variables, the yield surface of the material is a (hyper-)surface in a space of six dimensions, whose analytical expression has the form:

$$g(\sigma_1, \sigma_2, \sigma_3, \theta_1, \theta_2, \theta_3) = 0. \tag{1.3.6}$$

Here θ_1 and θ_2 are two additional angles that together with θ_3 help to define the principal stress directions with respect to the assumed co-ordinate axes. For yield curve (1.3.5) to be a two-dimensional planar cross-section of yield surface (1.3.6), the three variables σ_1, σ_2 and θ_3 appearing in its expression cannot be independent of each other, but must meet the condition that they should belong to the same two-dimensional plane in the six-dimensional space where surface g is. This would require that σ_1, σ_2 and θ_3 should be related to each other by a linear relation of the kind:

$$c_1\sigma_1 + c_2\sigma_2 + c_3\theta_3 = 0, \tag{1.3.7}$$

the coefficients c_i appearing here being independent of σ_1, σ_2 and θ_3. A simple glance at eqs (1.3.3) and (1.3.4), however, shows that far from satisfying eq. (1.3.7), the relation between σ_1, σ_2 and θ_3 is in the considered case highly non-linear. As a consequence, the yield curve (1.3.5) is not a two-dimensional planar cross-section of surface g. Therefore, it need not be convex, even if the latter is. Figure 1.3.1 provides a geometric representation of such a situation.

It should be noted, finally, that the above example does not hold true if the yield surface of the material is isotropic (which is usually the case for polycrystalline metals in virgin conditions). Isotropy of the yield surface requires that function g should not depend on stress direction. Accordingly, variables θ_1, θ_2 and θ_3 should be dropped from eq. (1.3.6). In particular, variable θ_3 should be dropped from eq. (1.3.5). In such a case, the considered (σ, τ)–yield curves would belong to the two-dimensional cross-section of g defined by plane σ_3 = constant. This would make them convex if surface g is such.

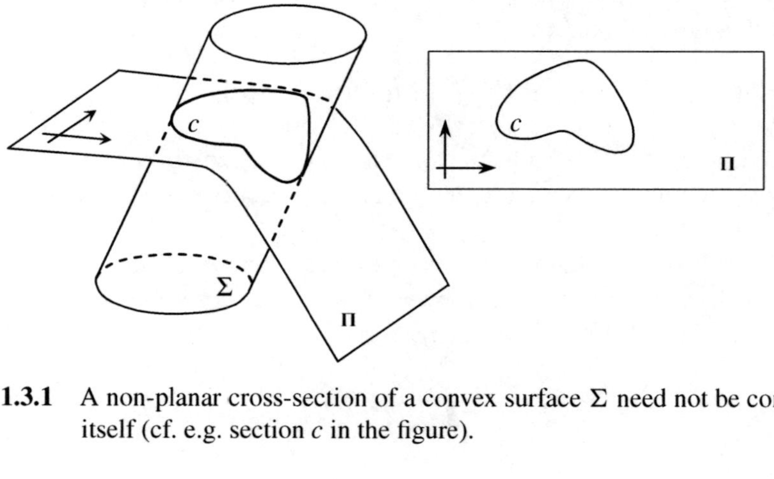

Fig. 1.3.1 A non-planar cross-section of a convex surface Σ need not be convex itself (cf. e.g. section *c* in the figure).

1.4 THE ROLE OF THE SUBSEQUENT YIELD SURFACES IN THE SOLUTION OF ELASTIC-PLASTIC BOUNDARY-VALUE PROBLEMS

The main purpose of this book is to establish a rational theory to determine the subsequent yield surfaces of a ductile metal, once its properties in the virgin state are known. This is a constitutive problem that can in principle be attacked independently of any consideration concerning equilibrium. The present Introduction could not be complete, however, without highlighting the role that the subsequent yield surfaces play in the solution of a typical equilibrium problem of an elastic-plastic material.

By and large, the final goal of Plasticity Theory is to determine the state of stress and strain (both elastic and plastic) of an elastic-plastic body subjected to a given loading history. This is achieved by solving an appropriate *boundary-value problem* relevant to the set of differential equations that govern the behaviour of the body when it is acted upon by surface tractions and displacements at its boundary. The set of governing equations may variously be composed, depending on the particular problem at hand and on the choice of the unknown variables. There are, however, two kinds of equations that cannot be left out of this set: the equilibrium equations and the plastic flow equations. If ρ denotes the mass density (per unit volume) and $\mathbf{b}$ the body force (force per unit mass), the equilibrium equations of any infinitesimal element of a material are given by:

$$\sigma_{ij,j} + \rho b_i = 0. \tag{1.4.1}$$

On the other hand, the plastic flow equations are nothing but the flow rule equations we introduced in Sect. 1.2. In their general form they are represented by eq. (1.2.10). If, as frequently happens in practice, the assumption of associate plastic flow is made, the same equations take the form (1.2.20), which also implies normality of the plastic flow direction with respect to the yield surface. In any case, knowledge of the yield function at any point of the material at any time of the process is needed in order to determine the plastic flow taking place in the material itself. In the case of eq. (1.2.10), the yield function enters the flow equations through the factor $\dot{\lambda}$, as specified by eq. (1.2.9). Whereas, in the case of eq. (1.2.20), the yield function enters the flow equations both through $\dot{\lambda}$ and explicitly through the term $\partial f/\partial \boldsymbol{\xi}^{(e)}$. Of course, this dependence of the boundary-value problem on the yield function is not at all surprising, as it is this function that controls elastic and plastic response at each point of the body.

Apart from the few fortunate cases where a closed form solution can be achieved analytically, the solution of an elastic-plastic boundary-value problem is a demanding task that can only be carried out numerically. The invaluable progress made in this field, however, has enormously simplified the work needed to carry out this task; so much so, that a solution of this problem is now within the reach of any structural engineer who can access one of the many commercially available computer codes. For a comparatively low price and in an incredibly short time, these codes afford a means to solve almost every elastic-plastic boundary-value problem with an astonishing degree of accuracy, no matter the complexity of the geometry of the body, the intricacy of its constitutive equations and the complications due to the occurrence of large deformations and displacements. Yet, still with the most sincere awe for the intellectual and practical value of these achievements, one cannot but remark that in order to produce realistic results all this computational power must be fed with realistic constitutive equations.

Here is where further work remains to be done. The currently available constitutive equations that regulate the evolution of the yield function of an elastic-plastic material are far from satisfactory compared with the available experimental data. Quite often, these equations reduce to a more or less complicated combination of the so-called kinematic and the isotropic work-hardening rules, which were originally proposed for their simplicity rather than for their adequacy to represent the experimental data in general. True enough, other work-hardening rules are also available, the slip plane theory being one of the possible alternatives. The range of applicability of all these rules, however, is quite limited and they can only provide realistic results for limited classes of processes and for particular states of stress. We have nowadays the power to solve any boundary-value problem of any elastic-plastic material for any conceivable loading history, but we lack general and realistic constitutive equations to model the evolution of the elastic range of the material as the inelastic deformation process proceeds. The chapters that follow aim at filling this gap for the wide class of ductile metals that fall under the name of von Mises materials.

2

Logical Premises to Subsequent Yielding

This chapter sets the ground for the deductive approach to subsequent yielding presented in this book. It contains a few elementary – even obvious – considerations. Some immediate conclusions will be drawn from them, the validity of which is hardly questionable. It will be shown from these simple premises that every subsequent yield condition of the material is implied by its yield condition in the virgin state.

2.1 DISTINGUISHING BETWEEN MATERIAL MATRIX AND MICROSCOPIC DEFECTS

It is well known that plastic deformation of ductile polycrystalline metals produces microscopic defects both in the crystal structure of the material and at the interface between the crystal grains. The defects can hardly be noticed on a macroscopic inspection since their dimension is in the atomic scale. Yet, their presence strongly affects some important macroscopic features of the material, such as yielding and work-hardening. All the microscopic defects we shall consider in this book are strictly those that are produced by plastic deformation. A material that does not contain such defects will be said to be defect-free or *virgin*, although it can – and in fact generally will – contain microscopic defects of a different origin.

The microscopic defects come in various kinds and sizes. Some of them are concentrated in a region that, from a macroscopic point of view, can, to all purposes, be assimilated to a geometric point. They are called *point defects*.

Other microscopic defects are spread over a line or a surface. Line and surface dislocations are the most common instances of these. The important point to be made here is that plastic deformation does not produce volume defect, that is, defects that occupy a macroscopic region of a non-vanishing volume. In other words, from the macroscopic standpoint, all microscopic defects produced by plastic deformation can be considered as volumeless. This is true if we exclude, as we shall henceforth, extremely large plastic deformations leading to void enucleation near rupture.

For the purpose of this book, it will be useful to regard the material as composed of two parts: the material that surrounds the defects, on the one hand, and the defects themselves, on the other. The material surrounding the defects will be called *matrix*.

An obvious consequence of this definition is that the material making up the matrix is defect free, that is, it is made of virgin material, whether the whole material is in a virgin state or not. Trivial though it may appear, this conclusion turns out to be fundamental to the analysis of this book.

In order to discriminate the defects from the surrounding matrix, we have to specify the portion of material making them up. The volume of this portion is negligible though, even when compared with the infinitesimal volume of the macroscopic element to which the defects belong. This is so because, as already said, all the defects considered in this book are volumeless when regarded from the macroscopic standpoint. We shall therefore somehow conventionally assume that the material making up the point, line and surface defects consists of the material that surrounds the defects a few atomic distances across. Since the defects are quite sparse and have geometrical dimensions less than three, the number of atoms making them up will be negligible when compared with the enormous number of atoms of the macroscopic volume element containing them. Therefore, though materialized in this way, the defects will still retain a negligible volume with respect to the infinitesimal volume element to which they belong. We are of course excluding pathological situations of unnaturally high defect concentration, resulting from extremely irregular deformation processes.

2.2 GEOMETRICAL AND PHYSICAL CONSEQUENCES

The lack of volume of the defects has several important consequences. The first one is the following:

Proposition 1. *The macroscopic deformation of any volume element of a material with defects coincides with the macroscopic deformation of its matrix.*

The reason for this is that the macroscopic deformation of an infinitesimal volume element of material can be viewed as the average of its deformation – possibly a highly irregular one – over the element volume. Since the defects occupy a negligible portion of the element, the deformation of the material making them up cannot significantly contribute to the mean deformation of the element. In the absence of volume defects, moreover, the volume of an element coincides with that of its matrix. Hence the above proposition.

Proposition 1 is tantamount to saying that the deformation of the whole material is the same as that of its matrix. The immediate consequence of this is:

Proposition 2. *A material with microscopic defects will yield plastically, if and only if its matrix will do so.*

But, being defined as the part of material that surrounds the defects, the matrix is defect-free itself. Or, to put it in another way, it is made entirely of virgin material. The inescapable conclusion of this and of Proposition 2 is then:

Proposition 3. *A cold worked material will yield plastically when its matrix will meet the same yield conditions that apply to the virgin material.*

Simple as its derivation is, this conclusion is of fundamental importance to the whole analysis of this book. It means that the subsequent yield conditions of a plastically deformed material can, at least in principle, be obtained by applying to its matrix the yield conditions that apply to the whole material in the virgin state. This may not be an easy task, though, since the matrix of a cold worked material is usually in a highly distorted state of microscopic strain due to the presence of microscopic defects. This state of microscopic strain, moreover, may not be known – or even knowable – precisely, since we may never be able to monitor the actual distribution and kind of defects produced by a deformation process. Yet Proposition 3 opens a new path in an otherwise dead-end situation. We shall exploit it in the following chapters in order to determine the subsequent yield surfaces of a wide class of elastic-plastic material of practical interest: the von Mises elastic-plastic materials.

2.3 ENTRAPPED ENERGY AND PERMANENT ELASTIC STRAIN

Before closing this chapter, it is appropriate to make some further considerations concerning the consequences of the presence of defects in a material.

Let us refer to an element of elastic-plastic material that has been set free from any macroscopic stress at its surface. In this condition, the element is said to be in a *stress-free state*. Though free from stress, the element will possess different amounts of elastic energy of microscopic distortion, depending on its previous plastic deformation process. In this state, moreover, the element will be differently deformed with respect to the stress-free state of the virgin material, depending on its previous plastic deformation process. The following discussion may help justifying the above claims.

The defects within the element apply a highly localized system of internal forces at the interface between the matrix and the defects. Being internal and composed of mutually opposed pairs, these forces do not require any macroscopic stress for the element to be in equilibrium. Yet they distort its atomic structure, thus making it store extra elastic energy at no macroscopic stress.

A part of this energy will remain in the element after it is brought back to its macroscopic stress-free state, because removing macroscopic stress from the surface of the element does not remove the defects it contains nor the microscopic distortion they produce. We shall call the elastic energy that remains in the element once all macroscopic stress is removed from it as *entrapped energy*. Its specific amount per unit mass will be denoted by $\Delta\psi$. It will of course depend on the previous plastic deformation process, since the number and kind of microscopic defects depend on it.

On the other hand, the microscopic elastic distortions due to the defects may result in a macroscopic deformation of the element of the material containing them. We shall define *permanent elastic strain* as the strain of the element, in its stress-free state, resulting from the microscopic elastic distortions due to the defects and measured by taking the stress-free state of the virgin material as reference configuration. Strictly speaking, permanent elastic strain is *elastic* in that it is due to the elastic microscopic distortions caused by the microscopic defects contained in the body. Yet it is *permanent*, since it remains in the material together with the defects as all macroscopic stress is removed from it. Though permanent, this strain should carefully be distinguished from plastic strain. The latter is permanent too, but it is not due to elastic microscopic deformation, since it is due to rigid-body sliding of some parts of the material with respect to others.

In the present book, we shall make frequent reference to a tensor $\boldsymbol{\varepsilon}^\circ$, which is opposite to permanent elastic strain as defined above. In terms of $\boldsymbol{\varepsilon}^\circ$, therefore, permanent elastic strain will be expressed as $-\boldsymbol{\varepsilon}^\circ$.

For many elastic-plastic materials, plastic deformation does not produce any volume change in the deformed body. This is a widely acknowledged experimental fact, which applies in particular to ductile metals. It means that the volume of an element of such a material in a macroscopic stress-free state is always the same and equal to the volume of the element in virgin conditions, no matter how much the element was plastically deformed. For these materials, it is reasonable to expect that permanent elastic strain should be represented by a traceless tensor; or, in other words, that tensor $\boldsymbol{\varepsilon}^{\circ}$ should be deviatoric. The reader should be warned, however, that such a hypothesis is not essential to the analysis presented in this book. Accordingly, it can easily be removed should experimental evidence turn out to be at variance with it.

As a final remark, it should be stressed that the possibility that the material may exhibit permanent strain as a result of the elastic microscopic distortions produced by microscopic defects is not usually recognized in Plasticity Theory. As a consequence, all deformation remaining in a material element once it is set free from macroscopic stress is reckoned as plastic. In fact, this deformation will in general contain the permanent elastic strain component, which entraps elastic energy and, thus, contributes to the total amount of elastic energy that is stored in the material. Permanent elastic strain will be shown to play a fundamental role in regulating the yield limit of many elastic-plastic metals.

3

Plastic Yielding under Deviatoric Energy Control

3.1 VON MISES ELASTIC-PLASTIC MATERIALS

Under isothermal conditions, most polycrystalline metals behave like isotropic linear elastic materials, provided that their stress or, equivalently, their elastic strain is kept within the yield surface. This is true whether the material is virgin or not. If the material is virgin, its yield limit is isotropic too. Accordingly, it depends on the principal values of stress but not on the principal stress directions. A vast amount of experimental evidence shows that the virgin yield surface of these materials is to a good approximation expressed by the Huber-von Mises yield criterion:

$$\psi_{\text{dev}} = \kappa^2. \tag{3.1.1}$$

Here, ψ_{dev} is the deviatoric part of the specific elastic energy stored in the material per unit volume. The reader can refer to Appendix A at the end of this book for a review of the most common expressions of specific elastic energy of linear elastic materials and of its deviatoric and volumetric parts. Constant κ^2 introduced in the above equation depends on the particular material under consideration. It represents the maximum amount of deviatoric strain energy that the material can store per unit volume at the considered temperature. If, as customary in Plasticity Theory, the yield stress in pure shear is denoted by k, then the strain energy at yield in pure shear is given by $k^2/2G$, where G is the elastic shear modulus. With this notation we have:

$$\kappa^2 = k^2/2G. \tag{3.1.2}$$

A more explicit expression of surface (3.1.1) can be obtained by substituting in its left-hand side any of the expressions of ψ_{dev} recalled in Appendix A, see eq. (A.27) or (A.29). In particular, we shall frequently refer to the expression of ψ_{dev} in terms of the principal values e_1^e, e_2^e and e_3^e of the elastic strain deviator $\boldsymbol{\varepsilon}^e$. This makes the yield surface of the virgin material assume the form:

$$\psi_{dev}(e_1^e, e_2^e, e_3^e) = \kappa^2 \tag{3.1.3}$$

or more explicitly, by recalling the equation in the last line of relation (A.27):

$$G \cdot (e_1^{e\,2} + e_2^{e\,2} + e_3^{e\,2}) = \kappa^2. \tag{3.1.4}$$

The analogous expression in terms of $\boldsymbol{\varepsilon}^e$, rather than $\mathbf{e}^e$, is readily obtained from eq. $(A.27)_4$:

$$\frac{1}{3}G[(\varepsilon_1^e - \varepsilon_2^e)^2 + (\varepsilon_2^e - \varepsilon_3^e)^2 + (\varepsilon_3^e - \varepsilon_1^e)^2] = \kappa^2. \tag{3.1.5}$$

It will be noticed that constant c°_{dev} appears in eq. (A.27) but not in eqs (3.1.4) and (3.1.5). This is due to the usual assumption – implicitly made here – that the elastic energy of a virgin material, and hence also its deviatoric component, is zero in the stress-free state. This makes $c^\circ_{dev} = 0$ for virgin materials. In eqs (3.1.3), (3.1.4) and (3.1.5), total deviatoric strains $\boldsymbol{\varepsilon}$ and $\mathbf{e}$ could equally well appear instead of the elastic deviatoric strains $\boldsymbol{\varepsilon}^e$ and $\mathbf{e}^e$, since in virgin materials elastic and total strains coincide.

Following a common practice, we shall name the yield surface defined by eq. (3.1.1) as *von Mises yield surface* . In fact, a yield surface of this kind was independently proposed by M.T. Huber in 1904 and R. von Mises in 1913. Before them, in a letter to W. Thompson dated 18 December 1856, J.C. Maxwell had expressed strong belief that plastic yielding begins as the deviatoric component of elastic energy reaches a certain limit, thus in essence implying a yield surface of the kind shown in eq. (3.1.1). Apparently, however, Maxwell did not pursue the argument any further and his letter was only published in 1936 (cf. [4, pp. 368–369]).

In order to keep the forthcoming analysis both simple and definite, we shall henceforth confine our attention to the particular but practically important class of von Mises elastic-plastic materials, defined as follows.

*A **von Mises material** is any elastic-plastic material satisfying the following requirements:*

1. *The virgin yield surface of the material can be expressed in the form (3.1.1).*
2. *When in the elastic range, the material responds as a linear elastic isotropic material, no matter the previous plastic deformation it suffered.*
3. *The elastic constants of the material are unaffected by plastic deformation.*

Requirements 1–3 of the above definition apply in good approximation to a wide range of metals of practical interest. Each of them can in principle be relaxed, though doing so is bound to complicate the analysis. Besides being well suited to many practical applications, the elastic-plastic theory of von Mises materials provides a suitable basis from which further generalization can be attempted.

It may be worth stressing that the above definition does in no way exclude that a cold worked von Mises material (i.e. a von Mises material in non-virgin condition) could exhibit anisotropic yield limits. In fact, the anisotropy of the yield limit due to plastic strain is the rule for these materials, taking place even after modest yielding. Yet, still exhibiting anisotropic yield limits as a result of cold work, the elastic behaviour of most elastic-plastic polycrystalline metals remains remarkably isotropic. Von Mises materials, as defined above, are perfectly adequate to model such behaviour.

3.2 ELASTIC ENERGY OF VON MISES MATERIALS

No matter whether it is virgin or not, a von Mises material is indistinguishable from a linear elastic isotropic material as long as its state of stress is kept in the elastic range. This means, in particular, that its elastic energy coincides with that of a linear elastic isotropic material, provided that (a) reference to elastic strain (rather than total strain) is made, and (b) the material is kept within its actual elastic region. The last condition can be dropped if the elastic properties of the material do not depend on plastic deformation. This applies to von Mises materials, according to requirement 3 of the above definition. Accordingly, the elastic energy of a von Mises material can be expressed in any of the forms (A.17), (A.18), (A.23) and (A.24) reported in Appendix A. In particular:

$$
\begin{aligned}
\psi &= \psi(\boldsymbol{\varepsilon}^{\mathrm{e}}) = \psi(\varepsilon_1^{\mathrm{e}}, \varepsilon_2^{\mathrm{e}}, \varepsilon_3^{\mathrm{e}}) \\
&= \frac{1}{2}\lambda(\varepsilon_1^{\mathrm{e}} + \varepsilon_2^{\mathrm{e}} + \varepsilon_3^{\mathrm{e}})^2 + \mu(\varepsilon_1^{\mathrm{e}\,2} + \varepsilon_2^{\mathrm{e}\,2} + \varepsilon_3^{\mathrm{e}\,2}) + c^{\circ}, \qquad (3.2.1)
\end{aligned}
$$

where c° is an appropriate constant [cf. eq. (A.23)]. For virgin materials $c^\circ = 0$, since, as already recalled, the elastic energy of the virgin material is taken to be equal to zero in the stress-free state. In virgin materials, moreover, total strain and elastic strain coincide. Thus $\boldsymbol{\varepsilon}$ could substitute $\boldsymbol{\varepsilon}^e$ in the above expression whenever reference to the virgin material is made.

In the forthcoming analysis, we shall have to take into account the effect of elastic distortions due to the microscopic defects that are produced by plastic deformation. As already discussed in Sect. 2.3, these distortions make the material store elastic energy at vanishing macroscopic stress. In general, therefore, a non-vanishing value of c° should appear in eq. (3.2.1) if this equation is referred to cold worked materials. This value depends on the elastic energy per unit volume that remains in the material once it recovers the stress-free – but in general, plastically deformed – state. Since in the stress-free state, the elastic energy of the virgin material is assumed to vanish, c° will coincide with the energy $\Delta\psi$ that is entrapped in the material as a result of the presence of microscopic defects produced by plastic deformation. That is:

$$\Delta\psi \equiv c^\circ. \tag{3.2.2}$$

In conclusion, as far as the elastic energy of the material is concerned, cold working a von Mises material can only produce a change in constant c° appearing in eq. (3.2.1) – the elastic constants of such a material being unaffected by plastic deformation. Geometrically speaking, this means that plastic deformation makes surface $\psi = \psi(\varepsilon_1^e, \varepsilon_2^e, \varepsilon_3^e)$ translate rigidly in the direction of the ψ-axis of the four-dimensional space $[\psi, \varepsilon_1^e, \varepsilon_2^e, \varepsilon_3^e]$ where it is defined. The amount of the translation is $\Delta\psi$. The situation is shown in Fig. 3.2.1.

3.3 SYMBOLIC EXPRESSION OF THE SUBSEQUENT YIELD SURFACES OF VON MISES MATERIALS

The determination of the yield surface of von Mises materials in virgin conditions is a straightforward task. We simply have to take the deviatoric part of their elastic energy ψ and insert it into the left-hand side of eq. (3.1.3). In fact, the task is even simpler in that we only need to know the value of the elastic shear modulus G of the material and introduce it in eq. (3.1.4). Of course, we must also fix the value of κ. The latter can be calculated through eq. (3.1.2) from the yield stress k in pure shear, or from the yield stress σ_y

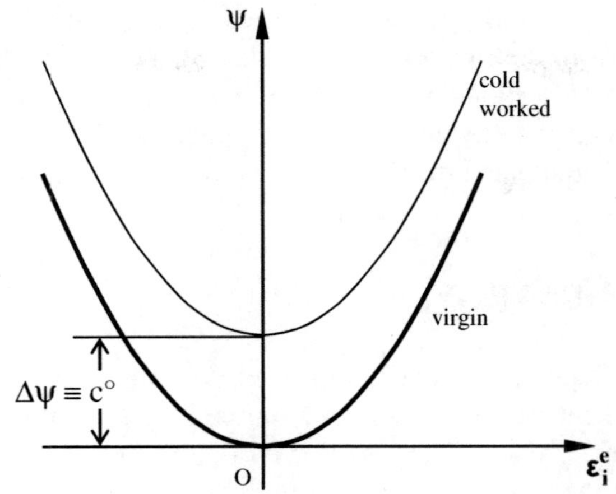

Fig. 3.2.1 A two-dimensional representation of the elastic energy surface of a von Mises material in virgin condition and after cold working, respectively. The curves are functions of the principal values of elastic strain and are meant to represent the actual elastic energy surfaces in four-dimensional space $[\psi, \varepsilon_1^e, \varepsilon_2^e, \varepsilon_3^e]$, cf. eq. (3.2.1).

in simple tension, since $k = \sigma_y/\sqrt{3}$, as is well known and implied by the yield condition (3.1.1) itself.

If the material is not virgin, condition (3.1.1) does not hold true, and the determination of the yield surface is not so immediate. According to Proposition 3 of Sect. 2.2, the yield surface of a von Mises material can be obtained by applying to its matrix the same yield condition (3.1.1) that applies to the virgin material. If $\overline{\psi}$ denotes the specific elastic energy of the material matrix per unit volume, the subsequent yield surfaces of a von Mises material are then given by:

$$\overline{\psi}_{dev} = \kappa^2. \qquad (3.3.1)$$

This expression reduces to eq. (3.1.1) in the case of virgin materials, since in this case $\overline{\psi} \equiv \psi$ and the matrix and the whole material are then one and the same thing.

Though quite general, yield criterion (3.3.1) is, as it stands, purely symbolic. In order to apply it to determine the subsequent yield surfaces of the material, we need to know beforehand the elastic energy $\overline{\psi}$ of its matrix. This does not appear to be an easy task, since the elastic energy stored in the matrix depends on the microscopic distortions that are produced by the microscopic defects embedded in it. The number, kind and location of the latter depends on the plastic process suffered by the material and can hardly be controlled from the macroscopic standpoint. Yet, as the analysis of the following chapters will show, the explicit expression of $\overline{\psi}$ can be obtained without much computational effort and in a purely deductive way. This will make eq. (3.3.1) quite an effective tool to predict and model the subsequent yield functions of any von Mises material.

3.4 DEFECT ENERGY

In closing this chapter, it may be worth observing that once the elastic energy of the matrix is introduced, it is only natural to consider the elastic energy of the whole material as composed of two parts: the elastic energy $\overline{\psi}$ of the matrix, on the one hand, and the remaining part ψ^*, on the other. In symbols:

$$\psi = \overline{\psi} + \psi^*. \tag{3.4.1}$$

As this definition should make clear, energy ψ^* is in fact a specific energy per unit volume, as are ψ and $\overline{\psi}$. It is a macroscopic continuous function of the macroscopic variables that describe the state of the material, since both ψ and $\overline{\psi}$ are such. We shall refer to ψ^* as the elastic specific energy stored in the defects per unit volume of the whole material, or *defect energy* for short. This definition makes sense in view of definition (3.4.1), considering that the whole material is supposed to be composed of two distinct phases (namely matrix and microscopic defects) and that, moreover, $\overline{\psi}$ is the matrix contribution to the energy ψ of the whole material.

As observed in Sect. 2.1, the microscopic defects we are considering are essentially volumeless. The energy ψ^* is, therefore, the macroscopic result from contributions that are highly singular even from the microscopic standpoint. Each contribution comes from a microscopic defect. It represents the work that would be produced or absorbed by the elastic forces exerted on the defect by the surrounding matrix should the material making up the defect be rearranged in such a way as to eliminate the defect itself. Clearly, this work depends on the macroscopic elastic strain at which such a rearrangement is supposed to occur, since the forces that surround the defects depend on it. Therefore, based on the value of the elastic strain of the embedding medium or, equivalently, of its stress, the contribution of each single defect to ψ can either be positive or negative. This means, in particular, that the sign of ψ^* may be different for different values of $\boldsymbol{\varepsilon}^e$. In this sense, the defect energy differs from ψ and $\overline{\psi}$, which are positive definite quantities, since they represent the elastic energy of a stable continuous material.

It should be apparent by now that, though formally identical to eq. (3.1.1), the general yield criterion (3.3.1) is in fact quite different from the former. As implied by eq. (3.4.1), the presence of defects in the material may make $\overline{\psi}$ differ from ψ, and thus the subsequent yield surfaces defined by eq. (3.3.1) differ from the virgin one. This is not surprising. In fact it is quite in agreement with the experimental evidence.

4

Geometric Representation of Strain and Strain Energy

This is admittedly the most demanding chapter of the book – but then dealing with non-isotropic functions of stress or strain has never been very easy. It introduces a special notation for second order symmetric tensors in terms of their principal values and principal directions. Though not intended for doing standard tensor algebra, the new notation lends itself to an intuitive geometrical interpretation that helps the analysis of the general geometrical structure of any isotropic scalar-valued functions of these tensors. This notation will accompany us through the rest of the book. It will be instrumental in establishing the main result of this chapter, represented by eq. (4.7.4). Thanks to this result, the derivation of the deviatoric energy of the matrix (Ch. 5) and the prediction of the subsequent yield surfaces of the material (Ch. 6) – whatever their anisotropy – will become a matter of course.

4.1 PRINCIPAL NOTATION FOR SYMMETRIC SECOND ORDER TENSORS

The arguments that follow are valid for any symmetric second order tensor. In presenting them, however, we shall make specific reference to tensor $\boldsymbol{\varepsilon}$, since we shall mainly apply them to strain tensors. A symmetric second order tensor can be assigned in a number of different ways. The standard way is to specify its components in a given coordinate system (x_1, x_2, x_3). This is the so-called component-form representation, and is the most suitable one to perform explicit tensor algebra. It amounts to specifying a set of six scalars, say $\{\varepsilon_{11}, \varepsilon_{22}, \varepsilon_{33}, \varepsilon_{12}, \varepsilon_{13}, \varepsilon_{23}\}$ or $\{\varepsilon_{ij}\}$ for short, representing the six independent components of $\boldsymbol{\varepsilon}$ in the considered coordinate system.

An alternative notation is to assign $\boldsymbol{\varepsilon}$ through its principal values $\varepsilon_1, \varepsilon_2$ and ε_3 together with the three Euler's angles ϕ, θ and ψ, defining the directions of the principal axes of $\boldsymbol{\varepsilon}$ with respect to the coordinate axes. In this case, tensor $\boldsymbol{\varepsilon}$ is defined by the set of six independent scalars $(\varepsilon_1, \varepsilon_2, \varepsilon_3, \phi, \theta, \psi)$. However, in view of the variety of different definitions of Euler's angles appearing in literature (see [5, p. 145]), we shall prefer to avoid using Euler's angles. Instead, we shall specify the principal directions of $\boldsymbol{\varepsilon}$ by means of an orthogonal tensor $\mathbf{Q}$ representing the rotation to be imposed to the coordinate axes x_1, x_2 and x_3 to superimpose them to the principal directions relevant to $\varepsilon_1, \varepsilon_2$ and ε_3, respectively. Tensor $\boldsymbol{\varepsilon}$ will thus be represented by the set of the three scalars $\varepsilon_1, \varepsilon_2$ and ε_3, and the orthogonal tensor $\mathbf{Q}$. This set will henceforth be denoted as $\{\varepsilon_1, \varepsilon_2, \varepsilon_3, \mathbf{Q}\}$ or, more succinctly, $\{\varepsilon_i, \mathbf{Q}\}$. Such a notation will be referred to as the *principal notation* and any tensor in such a notation will be said to be represented in *principal form*.

Set $\{\varepsilon_1, \varepsilon_2, \varepsilon_3, \mathbf{Q}\}$ is in fact a set of 12 scalars, since $\mathbf{Q}$ has nine scalar components. The latter, however, are not independent of each other, as they must satisfy the orthogonality conditions:

$$\mathbf{Q}\mathbf{Q}^{\mathsf{T}} = \mathbf{1} \quad \text{i.e. } Q_{ij}Q_{kj} = \delta_{ik}. \tag{4.1.1}$$

Only six of these conditions are independent, which reduces the number of independent components of $\mathbf{Q}$ to 3. Thus, also the principal-form representation defines a symmetric second order tensor in terms of six independent scalar quantities. Though introducing some redundancy, the use of $\mathbf{Q}$, instead of ϕ, θ and ψ, has the advantage of providing a more straightforward definition of rotation. The components of $\mathbf{Q}$ are easily calculated once it is remembered that its generic component Q_{ij} is the cosine of the angle that the principal axis *i* of $\boldsymbol{\varepsilon}$ forms with the coordinate axis x_j (see Appendix B).

Of course, if tensor $\boldsymbol{\varepsilon}$ is assigned in the component-form notation, its principal values and principal directions can be obtained by means of standard formulae (cf. e.g. [3, Sect. 4.2]). Conversely, if the principal-form notation of $\boldsymbol{\varepsilon}$ is adopted, a simple application of the transformation rule of tensor components under orthogonal coordinates transformation (cf. e.g. [5, Sect. 5.2]) enables us to revert to the component-form notation through the familiar formula:

$$\varepsilon_{ij} = Q_{ki}\varepsilon'_{kl}Q_{lj}. \tag{4.1.2}$$

Here ε'_{kl} are the components of $\boldsymbol{\varepsilon}$ in a coordinate system whose axes coincide with the principal directions of $\boldsymbol{\varepsilon}$. That is:

$$\varepsilon'_{kl} = \begin{bmatrix} \varepsilon_1 & 0 & 0 \\ 0 & \varepsilon_2 & 0 \\ 0 & 0 & \varepsilon_3 \end{bmatrix}. \tag{4.1.3}$$

The component-form notation and principal-form notation are, therefore, equivalent and the choice between them is simply a matter of convenience.

4.2 TWO GEOMETRIC INTERPRETATIONS OF THE PRINCIPAL NOTATION

The principal notation lends itself to two direct geometric interpretations, which will provide invaluable help in the course of our analysis. The first interpretation is a classical one and is based on Lamé ellipsoids. It will come in handy in Ch. 7, where the work-hardening law of the considered materials will be formulated. The second interpretation requires the introduction of special frames of reference. Though less straightforward than the former, this interpretation will provide a powerful tool for determining the general expression of the matrix strain energy.

4.2.1 Lamé ellipsoid representation

In space (x_1, x_2, x_3), tensor $\boldsymbol{\varepsilon}$ is represented by an ellipsoid whose axes are proportional to the principal values $\boldsymbol{\varepsilon}$ and directed as its principal directions (*Lamé ellipsoid*), Fig. 4.2.1a. The position of the ellipsoid centre is irrelevant as far as the present representation is concerned. However, in order to define the relative rotation of the principal directions of any two tensors, the axes of the ellipsoids that represent them must be labelled. For an easier pictorial representation, tensor $\boldsymbol{\varepsilon}$ is sometimes represented by an ellipse in a two-dimensional space, as in Fig. 4.2.1b. This is especially useful when dealing with plane tensors.

It may be worth recalling here that a symmetric second order tensor is invariant under a 180° rotation about any of its principal axes. This means that the ellipsoid that represents a tensor after such a rotation is indistinguishable from the one relevant to the unrotated tensor. The considered representation complies with this requirement. It must be pointed out, however, that as a consequence of its quadratic nature, the same ellipsoid corresponds to different tensors which differ only in the sign of some or all their

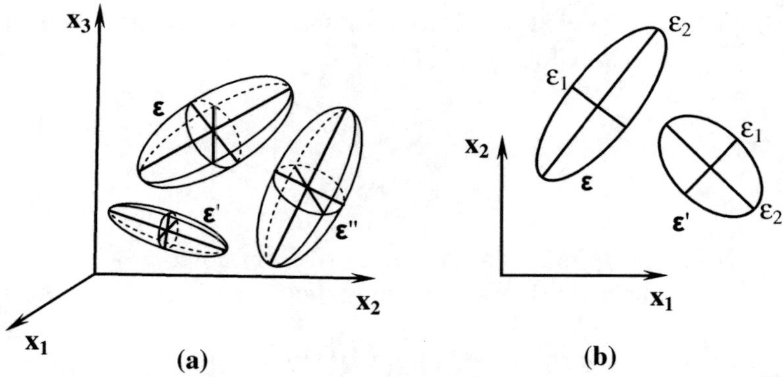

Fig. 4.2.1 Lamé ellipsoid representation. (a) Full spatial representation of symmetric second order tensors by means of ellipsoids. (b) Two-dimensional representation of the former.

principal values. When this difference matters, further information must be added in the ellipsoid representation to spot it. One way of doing this is to represent the ellipsoid axes by solid lines if they are relevant to positive principal values and by dashed lines otherwise.

Whether plane or spatial, Lamé representation is immediate when a tensor is given in principal-form notation. It should also be observed that the same notation enables us to easily distinguish between tensors, such as $\boldsymbol{\varepsilon}' \equiv \{\varepsilon_1, \varepsilon_2, \varepsilon_3, \mathbf{Q}'\}$ and $\boldsymbol{\varepsilon}'' \equiv \{\varepsilon_1, \varepsilon_2, \varepsilon_3, \mathbf{Q}''\}$, possessing the same principal values but different principal directions. In this case, Lamé representation produces two identical but differently oriented ellipsoids, which eloquently illustrates the similitude of the two tensors [cf. $\boldsymbol{\varepsilon}$ and $\boldsymbol{\varepsilon}''$ in Fig. 4.2.1a]. Such a similitude is hardly apparent when the same tensors are given in component-form notation, since the tensor components bear no obvious trace of it.

The second geometric interpretation of the principal-form representation of symmetric second order tensors is the following.

4.2.2 Rotated frame representation

Any symmetric second order tensor $\boldsymbol{\varepsilon}$ can be represented as a point of coordinates $[\varepsilon_1, \varepsilon_2, \varepsilon_3]$ in a rotated coordinate system (x'_1, x'_2, x'_3) whose axes are directed as the principal axes of $\boldsymbol{\varepsilon}$. The axes of (x'_1, x'_2, x'_3) are therefore rotated by $\mathbf{Q}$ with respect to those of the coordinate system (x_1, x_2, x_3), $\mathbf{Q}$ being the rotation tensor defined in Sect. 4.1. Again, the position of the origin of the rotated system does not matter. Such a representation establishes a one-to-one correspondence between symmetric second order tensors, on the one hand, and points in suitably rotated coordinate systems, on the other (see Fig. 4.2.2). It will accordingly be referred to as the *rotated frame representation*. Again, this geometric representation is immediate when a tensor is given in principal notation, since in this case both principal values and rotation tensor **Q** are explicitly specified.

This representation can be very useful when dealing with sets of tensors possessing the same principal directions. Any tensor within any such set is simply represented by its principal values and hence by a point or a position vector in the same rotated coordinate system (x'_1, x'_2, x'_3). It can easily be verified that if we want to add or subtract any two tensors of the set, or if we want to multiply one of them by a scalar, we can perform these operations on the corresponding position vectors in the rotated coordinate system in which they are represented. Although these operations are quite simple to be performed in component-form notation too, the considered representation gives them an immediate geometric meaning.

The rotated frame representation is also helpful in analysing material properties that depend upon a symmetric second order tensor in an isotropic way, i.e. independently of the directions of its principal axes. In this case, one can limit the attention to tensors possessing an arbitrarily fixed triad of principal axes, the actual direction of the latter being irrelevant since the considered material property is independent of it. Thus, any geometrical interpretation concerning that property can be made by referring to just one single rotated coordinate system. What is valid in that system will then hold

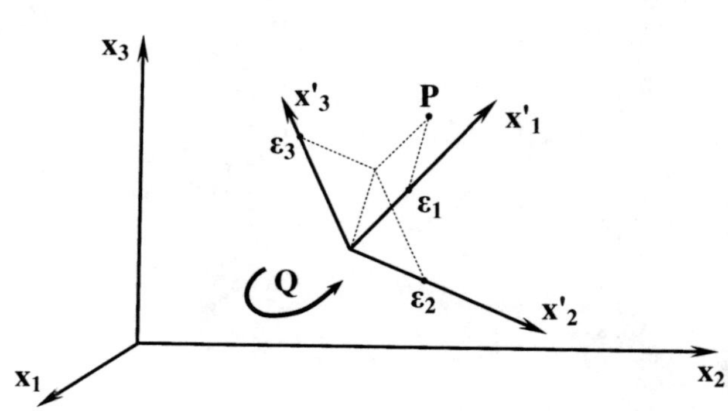

Fig. 4.2.2 Rotated frame representation: point $P \equiv [\varepsilon_1, \varepsilon_2, \varepsilon_3]$ in rotated coordinate system (x'_1, x'_2, x'_3) represents symmetric second order tensor $\boldsymbol{\varepsilon} \equiv \{\varepsilon_1, \varepsilon_2, \varepsilon_3, \mathbf{Q}\}$.

true in every rotated coordinate system; which may greatly simplify the analysis.

4.3 SCALAR-VALUED FUNCTIONS OF STRAIN

The rotated frame representation enables us to give an intuitive geometrical meaning to scalar-valued functions of symmetric second order tensors. To be definite, we shall refer to the strain energy function $\psi = \psi(\boldsymbol{\varepsilon})$. The same arguments apply, however, to any scalar-valued function of any symmetric second order tensor.

Let us first of all observe that if we express $\boldsymbol{\varepsilon}$ in principal notation we can write:

$$\psi = \psi(\boldsymbol{\varepsilon}) = \psi(\varepsilon_1, \varepsilon_2, \varepsilon_3, \mathbf{Q}). \qquad (4.3.1)$$

In rotated frame representation, tensor $\boldsymbol{\varepsilon}$ and hence the whole set of variables appearing at the far right-hand side of this equation are represented by a point in a suitably rotated coordinate system (x'_1, x'_2, x'_3). The axes of the latter are rotated by $\mathbf{Q}$ with respect to the triad of axes of the reference system (x_1, x_2, x_3). By adding a further independent dimension, say ψ, to (x'_1, x'_2, x'_3), we obtain a four-dimensional orthogonal coordinate system, namely $(\psi, \varepsilon_1, \varepsilon_2, \varepsilon_3)$, whose ε_1-, ε_2- and ε_3- axes coincide respectively with the x'_1-, x'_2- and x'_3- axes of the original coordinate system. The new coordinate system is represented in Fig. 4.3.1, where spatial dimension $x'_3 \equiv \varepsilon_3$ has been omitted for simplicity's sake.

In space $(\psi, \varepsilon_1, \varepsilon_2, \varepsilon_3)$, function (4.3.1) represents a family of surfaces (actually four-dimensional hyper-surfaces), each of which corresponds to

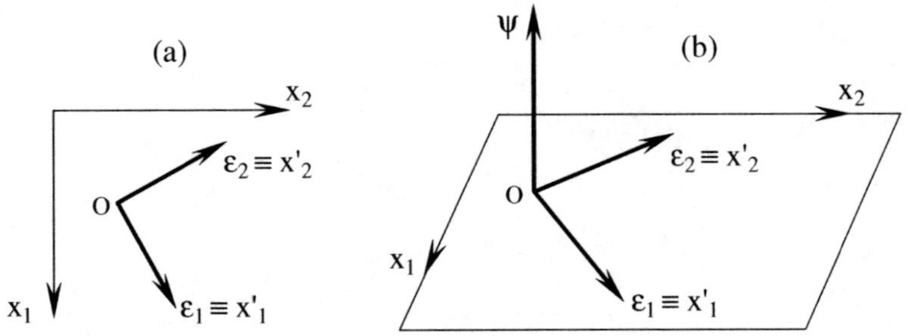

Fig. 4.3.1 An orthogonal axis ψ is added to the coordinate system $(\varepsilon_1, \varepsilon_2, \varepsilon_3) \equiv (x'_1, x'_2, x'_3)$ shown in (a). The resulting four-dimensional coordinate system $(\psi, \varepsilon_1, \varepsilon_2, \varepsilon_3)$ is shown in (b). (Axes x_3 and ε_3 are omitted from the picture.)

a different value of **Q** and hence to a different triad of principal directions for $\boldsymbol{\varepsilon}$. In rotated frame representation, therefore, each surface refers to a different coordinate system $(\psi, \varepsilon_1, \varepsilon_2, \varepsilon_3)$, the ε_1-, ε_2- and ε_3- axes of each system being rotated by **Q** with respect to the triad of axes $\{x_1, x_2, x_3\}$.

In general, the surfaces of this family will be different from one another – even when observed from the particular rotated coordinate systems $(\psi, \varepsilon_1, \varepsilon_2, \varepsilon_3)$ to which each surface belongs. The reason is that function (4.3.1) will in general associate different values of ψ to the given principal values $\varepsilon_1, \varepsilon_2$ and ε_3, depending on the value of **Q**, i.e. on the direction of the principal axes of $\boldsymbol{\varepsilon}$. In such a situation, function (4.3.1) is not invariant under rotation of the principal axes of $\boldsymbol{\varepsilon}$ or, equivalently, under rotation of the spatial reference coordinate system (x_1, x_2, x_3) in which it is defined. It must accordingly be classed as *anisotropic*.

Figure 4.3.2 shows two surfaces of family (4.3.1) in the rotated frame representation. It refers to the general case in which function $\psi = \psi(\boldsymbol{\varepsilon})$ is anisotropic. The figure shows how, in that representation, the lack of isotropy of this function will produce different surfaces ψ in different coordinate systems $(\psi, \varepsilon_1, \varepsilon_2, \varepsilon_3)$ relevant to different values of **Q**.

For particular forms of function $\psi = \psi(\boldsymbol{\varepsilon})$, it may well happen that the surfaces of the above family are the same for all tensors **Q** belonging to an appropriate group of orthogonal tensors. Function (4.3.1) is then said to be *invariant* under the orthogonal transformations belonging to that group. The latter is called the *symmetry group* of the function and taken as characterizing

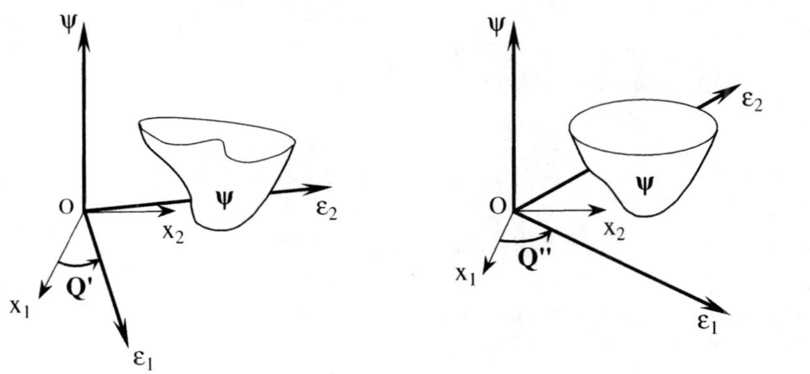

Fig. 4.3.2 Rotated frame representation of eq. (4.3.1) in the anisotropic case. Different ψ-surfaces correspond to different directions of the principal axes of $\boldsymbol{\varepsilon}$, i.e. to different rotations **Q** of the coordinate system $(\psi, \varepsilon_1, \varepsilon_2, \varepsilon_3)$.

its class of symmetry. If the symmetry group coincides with the full group of all the orthogonal tensors, function (4.3.1) is said to be *isotropic*.

Strictly speaking, one should distinguish between tensors $\mathbf{Q}$ belonging to the proper orthogonal group (Det $\mathbf{Q} = 1$) and those belonging to the full orthogonal group (Det $\mathbf{Q} = \pm 1$). Tensors of the former group are responsible for axis rotations, while those of the latter group also include axis inversions. When dealing with scalar-valued functions of second order tensors, however, the distinction between the two groups is irrelevant, since the law of transformation of the components of any even order tensor implies that their components remain unchanged under axis inversion. Thus, any scalar-valued function of a second order tensor is invariant under axis inversion and, accordingly, there is no point in distinguishing between the two groups in this case.

4.4 ISOTROPIC SCALAR-VALUED FUNCTIONS OF STRAIN

As already observed, a function $\psi = \psi(\boldsymbol{\varepsilon})$ is isotropic (i.e. invariant under all orthogonal transformations of the reference axes) if it associates the same value of ψ to the given tensor components $\{\varepsilon_{ij}\}$, no matter how the ε_i-axes are rotated about the ψ-axis. For fixed values of $\{\varepsilon_{ij}\}$, such a rotation of the coordinate system means a similar rotation of the principal directions of tensor $\boldsymbol{\varepsilon}$. It follows that isotropy implies that function $\psi = \psi(\boldsymbol{\varepsilon})$ should not depend on the principal directions of $\boldsymbol{\varepsilon}$.

If tensor $\boldsymbol{\varepsilon}$ is expressed in principal notation, this requirement can be imposed quite easily. We simply have to drop variable $\mathbf{Q}$ from eq. (4.3.1). Accordingly, the expression of an isotropic scalar-valued function of $\boldsymbol{\varepsilon}$ reduces to:

$$\psi = \psi(\boldsymbol{\varepsilon}) = \psi(\varepsilon_1, \varepsilon_2, \varepsilon_3). \tag{4.4.1}$$

Here, $\psi(\varepsilon_1, \varepsilon_2, \varepsilon_3)$ coincides with the function we get from the far right-hand side of eq. (4.3.1) once we assign $\mathbf{Q}$ a fixed value – any value will do, since the function is isotropic. Obviously, eq. (4.4.1) implies that ψ can also be expressed as a function of the principal invariants of $\boldsymbol{\varepsilon}$, namely $I_1(\boldsymbol{\varepsilon}), I_2(\boldsymbol{\varepsilon})$ and $I_3(\boldsymbol{\varepsilon})$, since the latter determine $\varepsilon_1, \varepsilon_2$ and ε_3 uniquely. In what follows, however, we shall not consider such a representation any longer.

Being independent of $\mathbf{Q}$, function (4.4.1) defines the *same* four-dimensional surface in *every* rotated coordinate system $(\psi, \varepsilon_1, \varepsilon_2, \varepsilon_3)$, no matter how the ε_i-axes are rotated or inverted. This means that as the ε_i-axes rotate about the ψ-axis, the surface that in the rotated system represents function (4.4.1) rotates rigidly with them. Moreover, if starting from a given placement of the ε_i-axes we invert the positive verse of one of them, say the ε_1-axis, the surface that represents $\psi = \psi(\boldsymbol{\varepsilon})$ in the inverted coordinate system must be the mirror image with respect to the (hyper-)plane $\varepsilon_1 = 0$ of the surface in the non-inverted system. Figure 4.4.1 illustrates the situation as far as the rotation of the ε_i-axes is concerned.

In rotated frame representation, therefore, the whole family of surfaces that are defined by the isotropic function (4.4.1) is entirely covered by any surface of the family. We just have to make this surface rotate rigidly about the ψ-axis and add the mirror images with respect to planes $\varepsilon_i = 0\,(i = 1, 2, 3)$ of the surfaces thus obtained.

4.5 SPECIAL ISOTROPIC FUNCTIONS

As observed at the end of the previous section, the family of surfaces that in the rotated frame representation represents an isotropic scalar-valued function of $\boldsymbol{\varepsilon}$ is quite specific. It can be generated entirely by any single surface of the family through rigid-body rotations and mirror reflections. Though strongly restricting the structure of the family, isotropy does not restrict shape, position or even symmetry of the surface that generates the family. In particular cases, of course, this surface can be endowed with particular geometric features and symmetries.

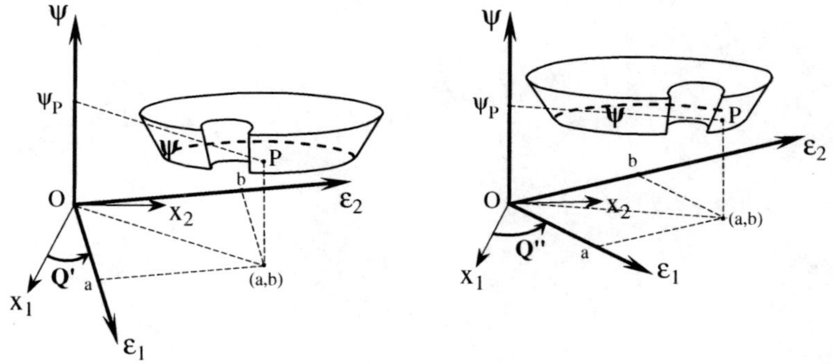

Fig. 4.4.1 Rotated frame representation of function $\psi = \psi(\boldsymbol{\varepsilon})$ in the isotropic case. The value ψ_P of ψ that corresponds to any given triplet of values for $\varepsilon_1, \varepsilon_2$ and ε_3 remains the same, no matter how the $(\psi, \varepsilon_1, \varepsilon_2, \varepsilon_3)$-coordinate system is rotated about the ψ-axis. This implies that surface ψ rotates rigidly with the ε_i-axes. Its shape is otherwise unrestricted.

Special symmetries are in fact possessed by the surfaces that represent many isotropic scalar-valued functions of practical interest. Take the elastic energy of a linear perfectly elastic material for instance. By observing that for a perfectly elastic material, total strain and elastic strain coincide and by recalling eq. (A.23) of Appendix A, we can write its expression in the form:

$$\psi = \psi(\varepsilon_1, \varepsilon_2, \varepsilon_3) = \frac{1}{2}\lambda(\varepsilon_1 + \varepsilon_2 + \varepsilon_3)^2 + \mu({\varepsilon_1}^2 + {\varepsilon_2}^2 + {\varepsilon_3}^2) + c^\circ, \quad (4.5.1)$$

λ, μ and c° being appropriate constants. It is not difficult to verify that in rotated frame representation the family of surfaces defined by function (4.5.1) is entirely made of (hyper-)paraboloids of elliptical cross-section. The axis of each paraboloid, moreover, coincides with the ψ-axis of the $(\psi, \varepsilon_1, \varepsilon_2, \varepsilon_3)$-coordinate system.

Two generic paraboloids of the family are shown in Fig. 4.5.1. A rotation $\mathbf{Q}$ of the principal axes of $\boldsymbol{\varepsilon}$ makes them rotate rigidly with the ε_i-axes. This suffices to assure isotropy of ψ. In addition to this, the paraboloids exhibit some obvious symmetries. The latter are a consequence of the additional property that the response of the material should be symmetric in tension and in compression. By this we mean, more specifically, that the same value of ψ should be obtained whether we apply eq. (4.5.1) to a point $[\varepsilon_1, \varepsilon_2, \varepsilon_3]$ or to its symmetric $[-\varepsilon_1, -\varepsilon_2, -\varepsilon_3]$ with respect to the origin of the ε_i-axes. It is not difficult to verify that eq. (4.5.1) embodies this property. In space

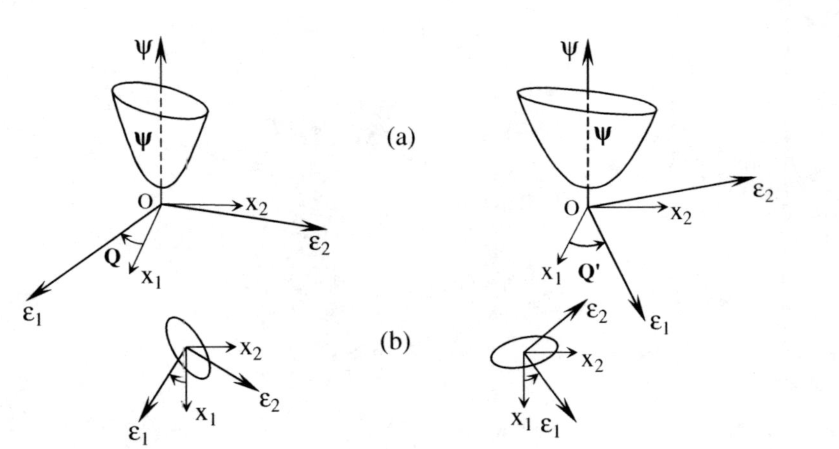

Fig. 4.5.1 (a) Two elliptic paraboloids of the family defined by eq. (4.5.1), relevant to two different values of $\mathbf{Q}$. (b) Their respective cross-sections at $\psi = \text{const}$.

$(\psi, \varepsilon_1, \varepsilon_2, \varepsilon_3)$, the paraboloid that represents this equation is, as a result, coaxial with the ψ-axis and endowed with elliptical cross-sections normal to that axis.

A still more particular shape is possessed by the deviatoric component ψ_{dev} of the elastic energy function (4.5.1). This is the scalar-valued function of $\boldsymbol{\varepsilon}$ given by [cf. Appendix A, eq. (A.27)]:

$$\psi_{dev} = \psi_{dev}(\varepsilon_1, \varepsilon_2, \varepsilon_3) = \frac{1}{3}G[(\varepsilon_1 - \varepsilon_2)^2 + (\varepsilon_2 - \varepsilon_3)^2 + (\varepsilon_3 - \varepsilon_1)^2] + c^\circ_{dev}, \tag{4.5.2}$$

where $G \equiv \mu$, while c°_{dev} is a further constant. In this case, any surface of the family is a (hyper-)paraboloid of *circular* cross-section, whose axis coincides with the ψ-axis of the coordinate system (Fig. 4.5.2). Being a surface of revolution about the ψ-axis, this paraboloid remains unaltered even with respect to a fixed coordinate system, such as (x_1, x_2, x_3), no matter how we rotate the ε_i-axes about the ψ-axis. Thus the whole family of surfaces reduces to just one single paraboloid of circular cross-section, fixed in space. Such a feature will come in handy in our quest for the matrix elastic energy of von Mises materials.

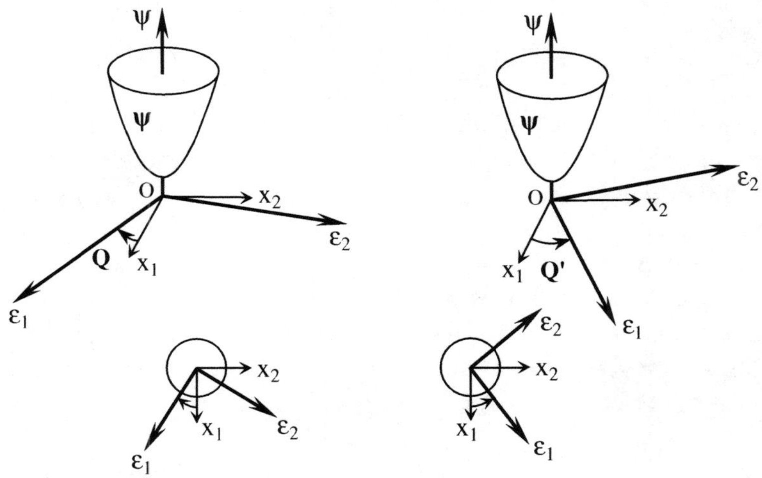

Fig. 4.5.2 The family of surfaces that represents the deviatoric part of the elastic energy of a linear elastic material is made of just one single paraboloid of revolution about the ψ–axis. The figure represents the paraboloid for two different placements of the rotating frame $(\psi, \varepsilon_1, \varepsilon_2, \varepsilon_3)$ and the relevant cross-sections at constant ψ.

4.6 CHANGE OF STRAIN REFERENCE CONFIGURATION

Any property of symmetry and isotropy of any function such as $\psi = \psi(\boldsymbol{\varepsilon})$ depends on the particular configuration (or reference state) of the material that is taken as a reference to measure strain. This is rather obvious, as one can hardly expect an otherwise elastic and isotropic material to exhibit an isotropic response to strain if the latter is measured from an anisotropically distorted reference state. In the present section we shall enquire how a change in the reference state modifies the analytical expression of a function such as $\psi = \psi(\boldsymbol{\varepsilon})$. The latter is assumed to be isotropic when the strain is measured from an appropriate reference configuration.

Let $\mathcal{R}^*$ be a given configuration of the material that we assume as reference configuration for strain, and let $\boldsymbol{\varepsilon}^*$ be the strain relevant to this configuration. We shall assume that ψ is isotropic in $\boldsymbol{\varepsilon}^*$. In view of eq. (4.4.1) this means that:

$$\psi = \psi(\varepsilon_1^*, \varepsilon_2^*, \varepsilon_3^*), \tag{4.6.1}$$

where ε_i^* are the principal values of $\boldsymbol{\varepsilon}^*$. Consider then another configuration of the material, say $\mathcal{R}$, and let $\boldsymbol{\varepsilon}$ be the strain tensor from this new configuration. If $\boldsymbol{\varepsilon}^\circ$ is the value of $\boldsymbol{\varepsilon}$ relevant to the deformation bringing the material from $\mathcal{R}$ to $\mathcal{R}^*$, the composition rule for small strain enables us to write:

$$\boldsymbol{\varepsilon} = \boldsymbol{\varepsilon}^\circ + \boldsymbol{\varepsilon}^*. \tag{4.6.2}$$

Though isotropic in $\boldsymbol{\varepsilon}^*$, function ψ will not in general be isotropic in $\boldsymbol{\varepsilon}$. As a matter of fact, if we exploit eq. (4.6.2) to express the principal values of $\boldsymbol{\varepsilon}^*$ as a function of the principal values of $\boldsymbol{\varepsilon}$, we shall generally end up with a function that also depends on the principal directions of $\boldsymbol{\varepsilon}$, and not simply on its principal values. This clearly destroys the isotropy of function ψ if the latter is expressed in the strain variable $\boldsymbol{\varepsilon}$.

The actual determination of ψ as a function of $\boldsymbol{\varepsilon}$ may require some tedious algebraic manipulations. However, this function can be obtained through straightforward geometric arguments in the particular, but practically important, case in which eq. (4.6.1) represents a surface of revolution about the ψ-axis.

In order to show how this can be done, let us start by giving tensors $\boldsymbol{\varepsilon}$ and $\boldsymbol{\varepsilon}^*$ a rotated frame representation. As already stated, $\boldsymbol{\varepsilon}$ and $\boldsymbol{\varepsilon}^*$ are two different strain measures since they refer to two different reference configurations, $\mathcal{R}$ and $\mathcal{R}^*$ respectively. It is natural, therefore, to represent these tensors in two different rotated frames. Accordingly, tensors $\boldsymbol{\varepsilon}$ will be represented by the points of a certain coordinate system, say $(\varepsilon_1, \varepsilon_2, \varepsilon_3)$, while tensors $\boldsymbol{\varepsilon}^*$ will be represented by the points of a different coordinate system, say $(\varepsilon_1^*, \varepsilon_2^*, \varepsilon_3^*)$. The origins of these coordinate systems will in general be different and denoted by O and O*, respectively.

The axes of one of the two coordinate systems can be directed arbitrarily, depending on the principal directions of the strain tensors that we mean to represent in them. However, if we want these two coordinate systems to represent strain tensors $\boldsymbol{\varepsilon}$ and $\boldsymbol{\varepsilon}^*$, respectively, possessing the same principal directions, we must take the triad of axes of the systems $(\varepsilon_1, \varepsilon_2, \varepsilon_3)$ and $(\varepsilon_1^*, \varepsilon_2^*, \varepsilon_3^*)$ parallel to each other. This ensues from the fact that in rotated frame representation, each point of the space spanned by the considered coordinate system represents a tensor whose principal directions are collinear with the coordinate axes. Which implies that the axes of the coordinate system $(\varepsilon_1, \varepsilon_2, \varepsilon_3)$ must parallel the homologous axes of the coordinate system $(\varepsilon_1^*, \varepsilon_2^*, \varepsilon_3^*)$ for the two systems to cover the same space of collinear tensors.

4.6.1 A particular case: $\boldsymbol{\varepsilon}$ and $\boldsymbol{\varepsilon}^*$ coaxial with $\boldsymbol{\varepsilon}^\circ$

Let us for the moment restrict our attention to the states of strain $\boldsymbol{\varepsilon}$ and $\boldsymbol{\varepsilon}^*$ that are coaxial with $\boldsymbol{\varepsilon}^\circ$. Let, moreover, the axes of both the coordinate systems $(\varepsilon_1, \varepsilon_2, \varepsilon_3)$ and $(\varepsilon_1^*, \varepsilon_2^*, \varepsilon_3^*)$ be collinear with the principal directions of $\mathbf{e}^\circ$. In these conditions, the origin O* of the coordinate system $(\varepsilon_1^*, \varepsilon_2^*, \varepsilon_3^*)$, if regarded as a point in the coordinate system $(\varepsilon_1, \varepsilon_2, \varepsilon_3)$, will represent the strain of configuration $\mathcal{R}^*$ with respect to $\mathcal{R}$; that is tensor $\boldsymbol{\varepsilon}^\circ$ itself. Accordingly, the location of point O* in the coordinate system $(\varepsilon_1, \varepsilon_2, \varepsilon_3)$ will be defined by position vector $\mathbf{p}^\circ \equiv \{\varepsilon_1^\circ, \varepsilon_2^\circ, \varepsilon_3^\circ\}$, where $\varepsilon_1^\circ, \varepsilon_2^\circ$ and ε_3° are the principal values of $\boldsymbol{\varepsilon}^\circ$. The coordinate systems $(\varepsilon_1, \varepsilon_2, \varepsilon_3)$ and $(\varepsilon_1^*, \varepsilon_2^*, \varepsilon_3^*)$ are shown in Fig. 4.6.1. For simplicity's sake, only two coordinate axes are represented in the picture.

In the same picture, rotation tensor $\mathbf{R}^\circ$ defines the directions of the principal axes of $\boldsymbol{\varepsilon}^\circ$ with respect to the triad of axes of the reference system (x_1, x_2, x_3). These are also the directions of the principal axes of $\boldsymbol{\varepsilon}$ and $\boldsymbol{\varepsilon}^*$, since tensors $\boldsymbol{\varepsilon}$, $\boldsymbol{\varepsilon}^*$ and $\boldsymbol{\varepsilon}^\circ$ are assumed to be coaxial. Tensors $\boldsymbol{\varepsilon}$ and $\boldsymbol{\varepsilon}^*$ will accordingly be represented by points in the reference system $(\varepsilon_1, \varepsilon_2, \varepsilon_3)$ and points in the reference system $(\varepsilon_1^*, \varepsilon_2^*, \varepsilon_3^*)$, respectively.

It should be noted that any point P in the space of the above coordinate systems represents a certain strain, say $\boldsymbol{\varepsilon}$, if it is considered in system $(\varepsilon_1, \varepsilon_2, \varepsilon_3)$, and quite another strain, say $\boldsymbol{\varepsilon}^*$, if it is regarded as belonging to system $(\varepsilon_1^*, \varepsilon_2^*, \varepsilon_3^*)$. These two different strains will in general represent two different states of deformation, measured from two different reference

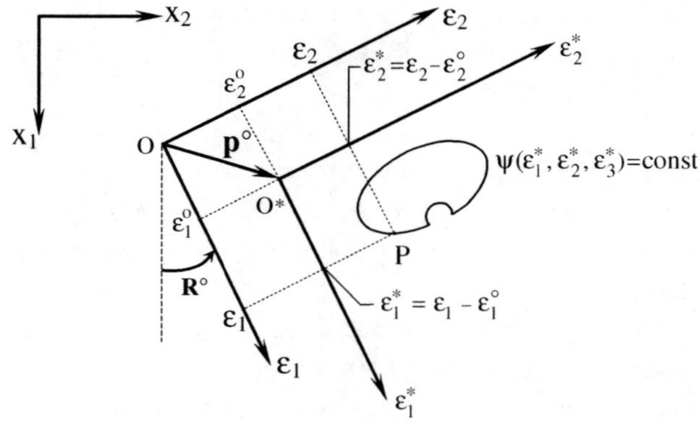

Fig. 4.6.1 Rotated frame representation of all tensors $\boldsymbol{\varepsilon}$ and $\boldsymbol{\varepsilon}^*$ that are coaxial with $\boldsymbol{\varepsilon}^\circ$ [the latter is represented by point O* in the coordinate system $(\varepsilon_1, \varepsilon_2, \varepsilon_3)$, the former are represented by the points of the coordinate systems $(\varepsilon_1, \varepsilon_2, \varepsilon_3)$ and $(\varepsilon_1^*, \varepsilon_2^*, \varepsilon_3^*)$, respectively]. A ψ = constant cross-section of surface (4.6.1) is also represented in the picture.

configurations ($\mathcal{R}$ and $\mathcal{R}^*$, respectively). They will not, therefore, be related to each other by eq. (4.6.2), as the latter refers to the *same* state of deformation measured from two different reference configurations of the material.

However, in the particular case in which the axes of the above coordinate systems are collinear with the principal axes of $\boldsymbol{\varepsilon}^\circ$, the three tensors $\boldsymbol{\varepsilon}$, $\boldsymbol{\varepsilon}^*$ and $\boldsymbol{\varepsilon}^\circ$ are coaxial. Their principal values will therefore be related by the following formula:

$$\varepsilon_i = \varepsilon_i^* + \varepsilon_i^\circ, \quad i = 1, 2, 3, \tag{4.6.3}$$

which is exactly the component-form expression of eq. (4.6.2) in principal components. Accordingly, the coordinates of P in the two coordinate systems $(\varepsilon_1, \varepsilon_2, \varepsilon_3)$ and $(\varepsilon_1^*, \varepsilon_2^*, \varepsilon_3^*)$ will in this case define the same state of deformation as measured from two different reference configurations, $\mathcal{R}$ and $\mathcal{R}^*$ respectively.

Of course, for a different direction of the coordinate axes, the tensors corresponding to the same point P in the two coordinate systems will still be collinear with each other, but not with $\boldsymbol{\varepsilon}^\circ$. This means that the difference $\boldsymbol{\varepsilon} - \boldsymbol{\varepsilon}^*$ cannot be collinear with $\boldsymbol{\varepsilon}^\circ$ and, therefore, $\boldsymbol{\varepsilon}$ and $\boldsymbol{\varepsilon}^*$ cannot meet eq. (4.6.3). Consequently, though corresponding to the same point P, tensors $\boldsymbol{\varepsilon}$ and $\boldsymbol{\varepsilon}^*$ will in general represent two different states of deformation, measured from the two different reference configurations $\mathcal{R}$ and $\mathcal{R}^*$, respectively.

Let us come back to the case of Fig. 4.6.1 in which $\boldsymbol{\varepsilon}$, $\boldsymbol{\varepsilon}^*$ and $\boldsymbol{\varepsilon}^\circ$ are coaxial. Let, moreover, $\psi(\varepsilon_1^*, \varepsilon_2^*, \varepsilon_3^*) = \text{const}$ be a generic cross-section of function (4.6.1) in the coordinate system $(\varepsilon_1^*, \varepsilon_2^*, \varepsilon_3^*)$, also reported in the same figure. As can easily be inferred from eq. (4.6.3), the analytical expression of this cross-section as a function of ε_1, ε_2 and ε_3 can be obtained from its expression as a function of ε_1^*, ε_2^* and ε_3^* by substituting $\varepsilon_1 - \varepsilon_1^\circ$, $\varepsilon_2 - \varepsilon_2^\circ$ and $\varepsilon_3 - \varepsilon_3^\circ$ for ε_1^*, ε_2^* and ε_3^*, respectively. This obviously means that for the subfamily of tensors $\boldsymbol{\varepsilon}$ that are coaxial with $\boldsymbol{\varepsilon}^\circ$, the expression of ψ in the coordinate system $(\varepsilon_1, \varepsilon_2, \varepsilon_3)$ will be given by:

$$\psi = \psi(\varepsilon_1 - \varepsilon_1^\circ, \varepsilon_2 - \varepsilon_2^\circ, \varepsilon_3 - \varepsilon_3^\circ), \tag{4.6.4}$$

where $\psi(\cdot, \cdot, \cdot)$ is the same function as that in (4.6.1).

As Fig. 4.6.1 should make it apparent, result (4.6.4) can also be obtained via a direct coordinate transformation from the coordinate system $(\varepsilon_1^*, \varepsilon_2^*, \varepsilon_3^*)$ to the coordinate system $(\varepsilon_1, \varepsilon_2, \varepsilon_3)$. As already observed, however, the validity of such a procedure and hence the validity of eq. (4.6.4) is strictly confined to the case in which tensors $\boldsymbol{\varepsilon}^*$ and $\boldsymbol{\varepsilon}^\circ$ are coaxial, because this also makes tensor $\boldsymbol{\varepsilon}$ coaxial with them, as implied by eq. (4.6.2).

4.6.2 The general case: principal axes of $\boldsymbol{\varepsilon}$ and $\boldsymbol{\varepsilon}^*$ however directed

When tensors $\boldsymbol{\varepsilon}$ and $\boldsymbol{\varepsilon}^*$ are not coaxial with $\boldsymbol{\varepsilon}^\circ$, eq. (4.6.4) is no longer valid, nor can the expression of ψ as a function of $\boldsymbol{\varepsilon}$ be obtained from function (4.6.1) via routine coordinate transformation. The reason for this is twofold. First of all, if $\boldsymbol{\varepsilon}$, $\boldsymbol{\varepsilon}^*$ and $\boldsymbol{\varepsilon}^\circ$ are not coaxial, the composition rule (4.6.2) does not imply a similar rule – such as eq. (4.6.3) – for the principal values of these tensors. Secondly, surface ψ moves as the ε_i^*-axes rotate (see Fig. 4.6.2). As a matter of fact this surface rotates rigidly with these axes, since the defining function (4.6.1) is assumed to be isotropic. As a consequence, coordinate transformation from system $(\varepsilon_1^*, \varepsilon_2^*, \varepsilon_3^*)$ to system $(\varepsilon_1, \varepsilon_2, \varepsilon_3)$ has to be applied to a surface whose position in the coordinate system $(\varepsilon_1, \varepsilon_2, \varepsilon_3)$ is different depending on the rotation suffered by these two systems. This makes its expression in terms of ε_1, ε_2 and ε_3 a matter that cannot be reduced to mere coordinate transformation.

For principal strains however directed, therefore, the explicit expression of $\psi(\boldsymbol{\varepsilon})$ as the reference configuration for strain is changed is, in general, far from immediate to obtain. It should also be emphasized that when the coordinate axes of $(\varepsilon_1, \varepsilon_2, \varepsilon_3)$ and $(\varepsilon_1^*, \varepsilon_2^*, \varepsilon_3^*)$ are not directed as the principal axes of $\boldsymbol{\varepsilon}^\circ$, point O* no longer represents tensor $\boldsymbol{\varepsilon}^\circ$ in the coordinate system

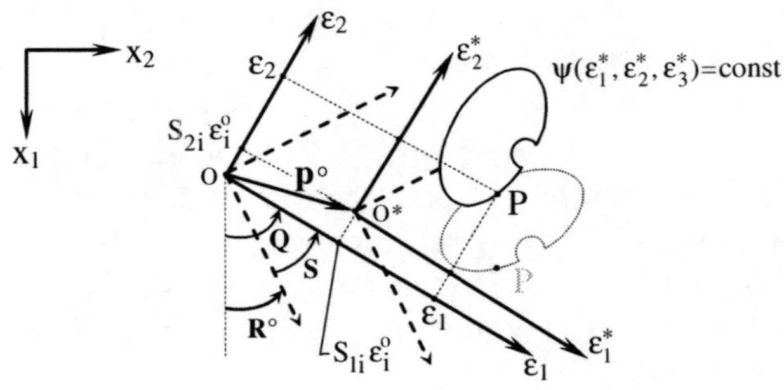

Fig. 4.6.2 Axes represented by dashed lines show the initial position of the coordinate systems $(\varepsilon_1, \varepsilon_2, \varepsilon_3)$ and $(\varepsilon_1^*, \varepsilon_2^*, \varepsilon_3^*)$. These systems suffer the same rotation **S** about their own origins O and O*, respectively. This modifies the relative position of the two coordinate systems (represented by solid lines axes). On the other hand, surface ψ, and thus its cross-sections $\psi(\varepsilon_1^*, \varepsilon_2^*, \varepsilon_3^*) =$ constant, rotates rigidly with the ε_i^*-axes since $\psi(\varepsilon_1^*, \varepsilon_2^*, \varepsilon_3^*)$ is assumed to be isotropic in ε_i^*. This changes the position of this surface – and hence its analytical expression – in the coordinate system $(\varepsilon_1, \varepsilon_2, \varepsilon_3)$, but not in the coordinate system $(\varepsilon_1^*, \varepsilon_2^*, \varepsilon_3^*)$.

$(\varepsilon_1, \varepsilon_2, \varepsilon_3)$. More precisely, if the axes of the coordinate system $(\varepsilon_1, \varepsilon_2, \varepsilon_3)$ are rotated by **S** with respect to the principal directions of tensor $\boldsymbol{\varepsilon}^\circ$, then point O*, or equivalently position vector $\mathbf{p}^\circ$, represents a tensor that is coaxial with the rotated coordinate axes. The principal values of this tensor are given by [cf. eq. (B.19)]:

$$p_i^\circ = S_{ij}\varepsilon_j^\circ . \tag{4.6.5}$$

This shows that the tensor represented by $\mathbf{p}^\circ$ is quite different from $\boldsymbol{\varepsilon}^\circ$, though it coincides with it for $\mathbf{S} \equiv \mathbf{1}$, i.e. when the two coordinate systems suffer no rotation.

4.7 ISOTROPY WITH CIRCULAR SYMMETRY ABOUT THE ψ-AXIS

An important special instance occurs when, in addition to being isotropic, function $\psi = \psi(\boldsymbol{\varepsilon}^*)$ represents a revolution surface about the ψ-axis (Fig. 4.7.1). In this case, in agreement with what we have already observed at the end of Sect. 4.5, surface ψ remains fixed in space $(\psi, \varepsilon_1, \varepsilon_2, \varepsilon_3)$, no matter how the ε_i^*-axes and the ε_i-axes are rotated. The invariance of this surface means, in particular, that for any rotation of the ε_i-axes, its analytical

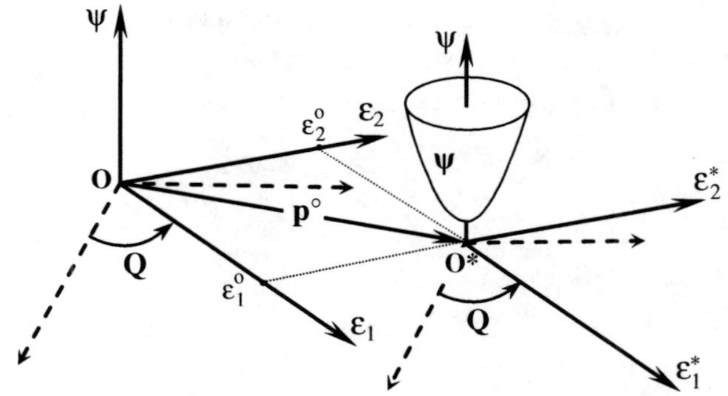

Fig. 4.7.1 Function $\psi = \psi(\varepsilon_1^*, \varepsilon_2^*, \varepsilon_3^*)$ is *both* isotropic in ε_i^* *and* represents a surface of revolution about the ψ-axis of the coordinate system $(\psi, \varepsilon_1^*, \varepsilon_2^*, \varepsilon_3^*)$. This makes it invariant under rotation $\mathbf{Q}$ of the coordinate systems $(\psi, \varepsilon_1, \varepsilon_2, \varepsilon_3)$ and $(\psi, \varepsilon_1^*, \varepsilon_2^*, e_3^*)$ about their respective ψ-axes.

expression in the rotated coordinate system $(\psi, \varepsilon_1, \varepsilon_2, \varepsilon_3)$ can be obtained through routine coordinate transformation from the expression of the same surface before rotation.

Suppose, in particular, that we want to determine the expression of surface ψ in the coordinate system $(\psi, \varepsilon_1, \varepsilon_2, \varepsilon_3)$ when the ε_i-axes are collinear with the ε_i°-axes (i.e. when the principal axes of tensor $\boldsymbol{\varepsilon}$ are collinear with the principal axes of tensor $\boldsymbol{\varepsilon}^\circ$). In this case, $\mathbf{S} = \mathbf{1}$ and eq. (4.6.4) applies. We can therefore write:

$$\psi = \psi(\varepsilon_1^*, \varepsilon_2^*, \varepsilon_3^*) = \psi(\varepsilon_1 - \varepsilon_1^\circ, \varepsilon_2 - \varepsilon_2^\circ, \varepsilon_3 - \varepsilon_3^\circ) = \widetilde{\psi}(\varepsilon_1, \varepsilon_2, \varepsilon_3). \quad (4.7.1)$$

Here, function $\widetilde{\psi}(\cdot,\cdot,\cdot)$ is obtained from function $\psi(\cdot,\cdot,\cdot)$ according to the following relation:

$$\widetilde{\psi}(\varepsilon_1, \varepsilon_2, \varepsilon_3) \equiv \psi(\varepsilon_1 - \varepsilon_1^\circ, \varepsilon_2 - \varepsilon_2^\circ, \varepsilon_3 - \varepsilon_3^\circ). \quad (4.7.2)$$

Thus, this is the expression of ψ as a function of $\boldsymbol{\varepsilon}$ in the coordinate system $(\psi, \varepsilon_1, \varepsilon_2, \varepsilon_3)$.

Let us now enquire how this expression of ψ would change as the coordinate systems $(\psi, \varepsilon_1, \varepsilon_2, \varepsilon_3)$ and $(\psi, \varepsilon_1^*, \varepsilon_2^*, \varepsilon_3^*)$ are simultaneously rotated by $\mathbf{S}$ about their ψ-axes. Since surface ψ is isotropic in $\boldsymbol{\varepsilon}^*$, a rotation of the coordinate system $(\psi, \varepsilon_1^*, \varepsilon_2^*, \varepsilon_3^*)$ will make this surface rotate about axis ψ rigidly with the rotating ε_i^*-axes. However, since ψ is also a revolution surface about the ψ-axis of the same coordinate system $(\psi, \varepsilon_1^*, \varepsilon_2^*, \varepsilon_3^*)$, such a rotation will leave it unaltered. It follows that its expression in any rotated coordinate system $(\psi, \varepsilon_1, \varepsilon_2, \varepsilon_3)$, say coordinate system $(\psi, \varepsilon'_1, \varepsilon'_2, \varepsilon'_3)$, can be obtained from function $\widetilde{\psi}(\varepsilon_1, \varepsilon_2, \varepsilon_3)$ as follows:

$$\psi = \widetilde{\psi}(S^T_{1j}\varepsilon'_j, S^T_{2j}\varepsilon'_j, S^T_{3j}\varepsilon'_j), \tag{4.7.3}$$

where we exploited the fact that the rotated coordinates ε'_i are related to coordinates ε_i by the well-known relation $\varepsilon'_i = S_{ij}\varepsilon_j$ [cf. eq. (B.16)].

Of course, rotation **S** does not affect the principal values of $\boldsymbol{\varepsilon}^\circ$, because they are invariant under coordinate rotation. From eqs (4.7.2) and (4.7.3) it then follows that the expression of surface ψ in the rotated coordinate system $(\psi, \varepsilon'_1, \varepsilon'_2, \varepsilon'_3)$ can be more specifically written as:

$$\psi = \psi(S^T_{1j}\varepsilon_j - \varepsilon^\circ_1, S^T_{2j}\varepsilon_j - \varepsilon^\circ_2, S^T_{3j}\varepsilon_j - \varepsilon^\circ_3). \tag{4.7.4}$$

In this equation, we wrote ε_i for ε'_i with the understanding that variables ε_i appearing there refer to the actual (rotated) coordinate system $(\psi, \varepsilon_1, \varepsilon_2, \varepsilon_3)$, previously referred to as $(\psi, \varepsilon'_1\varepsilon'_2\varepsilon'_3)$.

The importance of result (4.7.4) should not be underestimated. The expression of a scalar-valued function of strain, say $\psi = \psi(\boldsymbol{\varepsilon}^*)$, that is isotropic when strain is measured from a privileged reference state of the material may be quite simple and easy to obtain. On the other hand, the expression of the same function when strain is measured from a generic reference state is, usually, rather involved. Equation (4.7.4) enables us to find this expression at once in the particular, but important, case in which the isotropic function $\psi = \psi(\boldsymbol{\varepsilon}^*)$ is circular symmetric about the ψ-axis.

Again, the presence of **S** among the independent variables on the right-hand sides of eqs (4.7.3) and (4.7.4) indicates that, when expressed in a generic coordinate system, ψ will depend on rotation **Q**, i.e. on the orientation of the principal directions of $\boldsymbol{\varepsilon}$. This immediately results from the fact that:

$$\mathbf{Q} = \mathbf{S}\mathbf{R}^\circ, \tag{4.7.5}$$

which implies that:

$$\mathbf{S} = \mathbf{Q}\mathbf{R}^{\circ\,-1} = \mathbf{Q}\mathbf{R}^{\circ T}, \tag{4.7.6}$$

because $\mathbf{R}^\circ$ is orthogonal. Function ψ is not, therefore, isotropic in $\boldsymbol{\varepsilon}$, in spite of the fact that it is isotropic when expressed as a function of $\boldsymbol{\varepsilon}^*$.

4.8 HOW BODY ROTATION AFFECTS THE EXPRESSION OF ANISOTROPIC SCALAR-VALUED FUNCTIONS OF ELASTIC STRAIN OR STRESS

The analysis of this section will not be used in the remaining part of the book. It is included here for the sake of completeness, since it bears some formal resemblance with that of the previous section although it has quite a different meaning. The results to be presented here can be applied to account for the effect of body rotation on non-isotropic yield surfaces, once their expression for a given placement of the body is known. The uninterested reader may skip this section without prejudice to the understanding of the rest of the book.

As in the previous sections, the symbol $\boldsymbol{\varepsilon}$ will stand for elastic strain. If desired, therefore, the present analysis can immediately be rephrased in terms of stress rather than strain, simply by substituting $\boldsymbol{\sigma}$ for $\boldsymbol{\varepsilon}$.

In the previous section, we considered a scalar-valued function of strain that was isotropic provided that strain was measured from a given privileged reference configuration. Strain from that configuration was denoted by $\boldsymbol{\varepsilon}^*$. The function was then represented as a surface in privileged coordinate systems $(\psi, \varepsilon_1^*, \varepsilon_2^*, \varepsilon_3^*)$ and it was observed that it would rotate rigidly with the coordinate axes as the latter are rotated about the ψ-axis to span all possible states of strain $\boldsymbol{\varepsilon}^*$. We then considered the particular case in which that surface was also a revolution surface about the ψ-axis of the privileged system. In that case, we were able to determine its expression in more general coordinate systems $(\psi, \varepsilon_1, \varepsilon_2, \varepsilon_3)$. Although the latter were understood to be related by affine coordinate transformation to the coordinate systems $(\psi, \varepsilon_1^*, \varepsilon_2^*, \varepsilon_3^*)$, they referred to strain tensors $\boldsymbol{\varepsilon}$ that were measured from different configurations than the privileged one. This made the considered function lose its isotropic property when expressed in terms of $\boldsymbol{\varepsilon}$. The explicit expression of that function in the new coordinate systems was found to be obtainable through eq. (4.7.4).

A different question – though governed by similar algebra – is the following: suppose that a scalar function of strain, say $\psi(\boldsymbol{\varepsilon})$, represents a certain anisotropic property of the material, giving the values of ψ due to the application of strain starting from a given initial position of the material. In general, the assumed anisotropy means that

$$\psi(\boldsymbol{\varepsilon}) \neq \psi(\mathbf{Q}^{\mathsf{T}} \boldsymbol{\varepsilon}\, \mathbf{Q}), \tag{4.8.1}$$

$\mathbf{Q}$ being a proper orthogonal tensor. Leaving the coordinate system $(\psi, \varepsilon_1, \varepsilon_2, \varepsilon_3)$ fixed, we would like to determine the expression $\psi_s(\boldsymbol{\varepsilon})$ of the same quantity ψ, when the initial position of the body is rigidly rotated by $\mathbf{S}$.

We start by observing that for any given value of $\boldsymbol{\varepsilon}$, function $\psi_S(\boldsymbol{\varepsilon})$ must clearly return the same value of ψ that in the unrotated material the same function returns for a strain $\boldsymbol{\varepsilon}'$, which is obtained by applying to $\boldsymbol{\varepsilon}$ a rotation $\mathbf{S}^\mathsf{T}$, opposite to $\mathbf{S}$. The rotated strain is therefore $\boldsymbol{\varepsilon}' = \mathbf{S}\,\boldsymbol{\varepsilon}\,\mathbf{S}^\mathsf{T}$, as immediately results from the fact that any rotation $\mathbf{Q}$ transforms tensor $\boldsymbol{\varepsilon}$ into tensor $\mathbf{Q}^\mathsf{T}\boldsymbol{\varepsilon}\,\mathbf{Q}$. It can thus be concluded that:

$$\psi_S(\boldsymbol{\varepsilon}) = \psi(\mathbf{S}\,\boldsymbol{\varepsilon}\,\mathbf{S}^\mathsf{T}). \tag{4.8.2}$$

In simpler words, the values of ψ relevant to the rotated material are the same as the values of ψ in the unrotated material, provided that they are calculated for strain tensors that are oppositely rotated to the rotation of the material.

Lest eq. (4.8.2) should be wrongly interpreted as resulting from a coordinate transformation, let us see what happens if we apply it to the case in which ψ is a function of $\boldsymbol{\varepsilon} - \boldsymbol{\varepsilon}^\circ$, where $\boldsymbol{\varepsilon}^\circ$ is any constant tensor. In this case, we have $\psi = \psi(\boldsymbol{\varepsilon} - \boldsymbol{\varepsilon}^\circ)$, and eq. (4.8.2) yields:

$$\psi_S(\boldsymbol{\varepsilon} - \boldsymbol{\varepsilon}^\circ) = \psi(\mathbf{S}\,\boldsymbol{\varepsilon}\,\mathbf{S}^\mathsf{T} - \boldsymbol{\varepsilon}^\circ). \tag{4.8.3}$$

It should be observed here that a rotation $\mathbf{S}$ of the coordinate system $(\psi, \varepsilon_1, \varepsilon_2, \varepsilon_3)$ would produce quite a different result. Such a rotation would transform function $\psi = \psi(\boldsymbol{\varepsilon} - \boldsymbol{\varepsilon}^\circ)$ in the unrotated coordinate system into function $\psi = \psi[\mathbf{S}^\mathsf{T}(\boldsymbol{\varepsilon} - \boldsymbol{\varepsilon}^\circ)\mathbf{S}] = \psi(\mathbf{S}^\mathsf{T}\,\boldsymbol{\varepsilon}\,\mathbf{S} - \mathbf{S}^\mathsf{T}\,\boldsymbol{\varepsilon}^\circ\,\mathbf{S})$ in the rotated one.

4.8.1 An application to kinematic hardening

To illustrate how eq. (4.8.2) can be put to practical use, let us consider the case of the so-called kinematic hardening. In this case, an initially isotropic yield surface $f(\boldsymbol{\sigma})$ is supposed to transform to $f(\boldsymbol{\sigma} - \boldsymbol{\rho})$ due to plastic straining. Here $\boldsymbol{\rho}$ is a second order symmetric tensor, usually referred to as *back-stress* (cf. e.g. [6, Sect. 3.3.5]).

By producing a translation of the yield surface in stress space, kinematic hardening destroys its initial isotropy, thus making the translated surface depend on body rotation. Since no rotation appears among the independent variables of $f(\boldsymbol{\sigma} - \boldsymbol{\rho})$, this expression must be intended to be valid for a particular placement of the body, or perhaps when the plastic deformation process does not produce any rotation at all in the material. Fundamental though this point is, it is hardly mentioned in the literature on this topic.

A possible reason may be that function $f(\boldsymbol{\sigma} - \boldsymbol{\rho})$ is tacitly assumed to be valid for small rotations – a rather restrictive assumption in the presence of plastic motion.

Equation (4.8.3) enables us to find the correct expression of the new yield surface at once. For instance, assuming that function $f(\boldsymbol{\sigma} - \boldsymbol{\rho})$ applies in the absence of any rotation, its correct expression when the body suffers a generic rotation **S**, is given by:

$$f_s(\boldsymbol{\sigma} - \boldsymbol{\rho}) = f(\mathbf{S}\,\boldsymbol{\sigma}\,\mathbf{S}^{\mathsf{T}} - \boldsymbol{\rho}), \tag{4.8.4}$$

according to eq. (4.8.3). This should be considered the correct general expression for kinematic hardening.

Whether kinematic hardening is at all adequate to describe the subsequent yield surfaces of a von Mises material is quite another question. The available experimental evidence confirms that such work-hardening can only be acceptable for special kinds of plastic deformation processes and then when sufficiently near the virgin surface.

5

The Elastic Energy of the Matrix

5.1 DETERMINING THE ELASTIC ENERGY STORED IN THE MATRIX

The specific elastic energy of the matrix of an elastic-plastic material with defects was introduced in Sect. 3.3 and denoted by $\overline{\psi}$. In a virgin material, there are no defects and, as observed in that section, $\overline{\psi} \equiv \psi$. From eq. (3.2.1) it then follows that the specific elastic energy of the matrix of a virgin material can be expressed as:

$$\overline{\psi} = \overline{\psi}^{\circ}(\boldsymbol{\varepsilon}) = \overline{\psi}^{\circ}(\varepsilon_1, \varepsilon_2, \varepsilon_3) = \frac{1}{2}\lambda(\varepsilon_1 + \varepsilon_2 + \varepsilon_3)^2 + \mu({\varepsilon_1}^2 + {\varepsilon_2}^2 + {\varepsilon_3}^2). \tag{5.1.1}$$

Here, tensor $\boldsymbol{\varepsilon}$ is the total strain from the macroscopic stress-free state of the material. It replaces tensor $\boldsymbol{\varepsilon}^{e}$ appearing in eq. (3.2.1), since in virgin materials elastic and total strain coincide. In writing eq. (5.1.1), we also set $c^{\circ} = 0$, since we assume that the elastic energy of a virgin material vanishes at the stress-free state.

Of course, the validity of eq. (5.1.1) is limited to virgin materials. If the material is not virgin, the elastic distortions due to microscopic defects can make $\overline{\psi}$ quite a different function of $\boldsymbol{\varepsilon}$, depending on the previous plastic deformation history of the material.

In order to determine the expression of the elastic energy of the matrix in non-virgin conditions, let us consider an infinitesimal volume element of a material in its macroscopic stress-free state after cold working. Let us then imagine that we remove all the microscopic defects from the element, possibly by cutting it as needed and welding it again once the defects are eliminated. The process is assumed to remove all the distorting actions that the defects exerted upon the surrounding matrix.

The latter will then recoil spontaneously to its fully relaxed stress-free state, since its energy is at a minimum at that state. This is a direct consequence of the fact that in the elastic range, a von Mises material behaves as a linear elastic material, which as we know possesses a stable equilibrium state in stress-free conditions.

Although such a relaxing process is usually far too complicated to be done in practice, it can sometimes be realized by making the element undergo an appropriate thermal treatment (annealing). For the purpose of the present analysis, however, we only need to recognize the sheer possibility of it for the class of materials we are considering. Clearly, the process would liberate the elastic energy $\Delta\overline{\psi}$ that is entrapped in the matrix due to the microscopic distortions that are produced by the defects. The following equation relates this energy to the energy $\Delta\psi \equiv c^\circ$ we introduced in Sect. 3.2:

$$\Delta\psi = \Delta\overline{\psi} + \Delta\psi^*, \tag{5.1.2}$$

as immediately follows from eq. (3.4.1). Energy $\Delta\psi^*$ appearing here is the contribution to $\Delta\psi$ from the defects when the material as a whole is in its macroscopic stress-free (but in general, microscopically distorted) state.

As anticipated in Sect. 2.3, the relaxing process considered above will generally bring the matrix to a state of macroscopic strain $\boldsymbol{\varepsilon}^\circ$ with respect to the stress-free state of the material with defects. The opposite strain, namely $-\boldsymbol{\varepsilon}^\circ$, is the strain of the material resulting from the microscopic distortions imposed on it by the defects. This is the permanent elastic strain we mentioned in Sect. 2.3. Since the matrix is virgin material, its strain energy function after the relaxing process will be expressed by eq. (5.1.1), provided that the strain tensor appearing there is measured from its new stress-free state. The latter is deformed by $\boldsymbol{\varepsilon}^\circ$ with respect to the original stress-free state of the material with defects. We shall denote the strain of the material with defects with respect to the fully relaxed stress-free state of the matrix by $\boldsymbol{\varepsilon}^*$. We clearly have that:

$$\boldsymbol{\varepsilon}^* = \boldsymbol{\varepsilon} - \boldsymbol{\varepsilon}^\circ, \tag{5.1.3}$$

in agreement with the strain composition rule for small strains.

As the above discussion should make apparent, the elastic energy $\overline{\psi}$ possessed by the matrix of a material with defects at any given state of strain can be considered as produced by two contributions. One contribution comes from the energy of the virgin material at the considered state of strain, measured starting from the fully relaxed state of the matrix. The other contribution is the part of $\Delta\overline{\psi}$ that the matrix would release at constant strain during the defect removal process we mentioned above. This part of $\Delta\overline{\psi}$ will be referred to as $\overline{c}$ – a constant that will clearly depend on the defect content of the material. By taking tensor $\boldsymbol{\varepsilon}^*$ as a measure of strain, we can, accordingly, express $\overline{\psi}$ in the following form:

$$\overline{\psi} = \overline{\psi}(\boldsymbol{\varepsilon}^*) = \overline{\psi}(\varepsilon_1^*, \varepsilon_2^*, \varepsilon_3^*) = \frac{1}{2}\lambda(\varepsilon_1^* + \varepsilon_2^* + \varepsilon_3^*)^2 + \mu(\varepsilon_1^{*2} + \varepsilon_2^{*2} + \varepsilon_3^{*2}) + \overline{c}. \tag{5.1.4}$$

Though perfectly correct and quite general, this expression can hardly be directly applied in practice to calculate the matrix elastic energy as a function of the total strain $\boldsymbol{\varepsilon}$ from the macroscopic stress-free state of the plastically deformed material, eq. (5.1.3). The reason is two-fold. First of all, the values of $\boldsymbol{\varepsilon}^\circ$ and $\overline{c}$ are usually unknown at the outset. Secondly, simple as it appears, eq. (5.1.4) by no way implies an analogous equation in terms of the differences $\varepsilon_i - \varepsilon_i^\circ$ between the principal values of $\boldsymbol{\varepsilon}$ and $\boldsymbol{\varepsilon}^\circ$. This is due to the fact that, though related by eq. (5.1.3), tensors $\boldsymbol{\varepsilon}^*$, $\boldsymbol{\varepsilon}$ and $\boldsymbol{\varepsilon}^\circ$ will not in general be coaxial with each other. It follows that, when expressed as a function of $\boldsymbol{\varepsilon}$ rather than $\boldsymbol{\varepsilon}^*$, the elastic energy of the matrix would lose its simple appearance (5.1.4) to assume a different and usually rather awkward form.

Yet, we need to express $\overline{\psi}$ in terms of $\boldsymbol{\varepsilon}$, since this is the strain measure that is most readily accessible to experiment and, as such, the most suitable one to describe the stress/strain behaviour of the material.

In the next section, we shall show how an important part of $\overline{\psi}$ can be given a rigorous and reasonably simple expression in terms of $\boldsymbol{\varepsilon}$. Though not representing the entire elastic energy stored in the matrix, this part of $\overline{\psi}$ does nonetheless contain all the information we need for a complete description of the yield surface of the elastic-plastic materials under consideration.

5.2 THE REDUCED ELASTIC ENERGY OF THE MATRIX

Rather than considering the entire elastic energy of the matrix, we shall henceforth confine most of the following considerations to the sole part of it that depends on the shear modulus of the material (variously indicated as μ or G in the available literature and in the present book). This part of the matrix elastic energy will be referred to as the *reduced elastic energy of the matrix*, and its specific value per unit volume denoted by $\overline{\psi}_\mu$. Its expression as a function of $\boldsymbol{\varepsilon}^*$ is taken as coinciding with the second term at the right-hand side of eq. (5.1.4). Moreover, only the deviatoric part of $\overline{c}$ is included in it. That is:

$$\overline{\psi}_\mu = \overline{\psi}_\mu(\varepsilon^*) = \overline{\psi}_\mu(\varepsilon_1^*, \varepsilon_2^*, \varepsilon_3^*) = \mu(\varepsilon_1^{*2} + \varepsilon_2^{*2} + \varepsilon_3^{*2}) + d_o, \qquad (5.2.1)$$

where d_o is the deviatoric part of $\overline{c}$.

The importance of $\overline{\psi}_\mu$ as far as the present analysis is concerned, stems from the fact that its deviatoric part coincides with the deviatoric part of the elastic energy of the matrix. That is:

$$\overline{\psi}_{dev} \equiv (\overline{\psi}_\mu)_{dev}. \qquad (5.2.2)$$

This equation can readily be verified once it is recalled that the deviatoric part of any scalar-valued function of $\boldsymbol{\varepsilon}^*$ is obtained by calculating the same function for the deviatoric component of $\boldsymbol{\varepsilon}^*$ and subtracting from it the volumetric part of its value for $\boldsymbol{\varepsilon}^* = \mathbf{0}$. Since the deviatoric component of a tensor is traceless, the first term on the right-hand side of eq. (5.1.4) vanishes when it is calculated for a deviatoric tensor. On the other hand, the deviatoric part of $\overline{c}$ is d_o, which fully justifies eq. (5.2.2).

Both $\overline{\psi}(\boldsymbol{\varepsilon}^*)$ and $\overline{\psi}_\mu(\boldsymbol{\varepsilon}^*)$ are isotropic functions of $\boldsymbol{\varepsilon}^*$. This is immediately apparent from the fact that their defining equations – (5.1.4) and (5.2.1), respectively – are independent of the principal directions of $\boldsymbol{\varepsilon}^*$. Function $\overline{\psi}_\mu(\boldsymbol{\varepsilon}^*)$, moreover, represents a (hyper-)surface of revolution about the $\overline{\psi}_\mu$-axis in the four-dimensional space defined by the co-ordinate system $(\overline{\psi}_\mu, \varepsilon_1^*, \varepsilon_2^*, \varepsilon_3^*)$. The latter property can easily be verified by observing that the cross-sections $\overline{\psi}_\mu = \text{const}$ of the surface defined by eq. (5.2.1) are (hyper-)circles centred at the intersection of the (hyper-) plane $\overline{\psi}_\mu = \text{const}$ with the axis $\overline{\psi}_\mu$. (In the reduced three-dimensional space $(\overline{\psi}_\mu, \varepsilon_1^*, \varepsilon_2^*)$, these cross-sections appear as ordinary circles, lying on the ordinary plane $\overline{\psi}_\mu = \text{const}$ and centred at the intersection of this plane with the $\overline{\psi}_\mu$-axis.)

Taking into account the axial-symmetry property of $\overline{\psi}_\mu$, we can apply result (4.7.4) to eq. $(5.2.1)_2$ and obtain the expression of $\overline{\psi}_\mu$ as a function of $\boldsymbol{\varepsilon}$ rather than of $\boldsymbol{\varepsilon}^*$. We thus get:

$$\overline{\psi}_\mu = \overline{\psi}_\mu(\boldsymbol{\varepsilon}) = \overline{\psi}_\mu(S_{1j}^T\varepsilon_j - \varepsilon_1^o, S_{2j}^T\varepsilon_j - \varepsilon_2^o, S_{3j}^T\varepsilon_j - \varepsilon_3^o). \qquad (5.2.3)$$

Here, ε_i° are the principal values of tensor $\boldsymbol{\varepsilon}^\circ$ introduced above. They depend on the plastic deformation history of the material.

A more explicit expression of $\overline{\psi}_\mu(\boldsymbol{\varepsilon})$ can be obtained from eq. (5.2.1)$_3$ once the substitution of variables implied by eq. (5.2.3)$_2$ is worked out explicitly. Since relation $\mathbf{S}\mathbf{S}^\mathsf{T} = \mathbf{1}$ implies that $S^\mathsf{T}_{ir}S^\mathsf{T}_{ik} = \delta_{rk}$, after some simple algebra we finally obtain:

$$\begin{aligned}\overline{\psi}_\mu &= \overline{\psi}_\mu(\boldsymbol{\varepsilon}) = \overline{\psi}_\mu(\varepsilon_1, \varepsilon_2, \varepsilon_3, \mathbf{S}) = \mu(\varepsilon_1{}^2 + \varepsilon_2{}^2 + \varepsilon_3{}^2) + a_i\, S^\mathsf{T}_{ij}\varepsilon_j + b \\ &= \mu(\varepsilon_1{}^2 + \varepsilon_2{}^2 + \varepsilon_3{}^2) + a_i\, S_{ji}\varepsilon_j + b, \qquad (5.2.4)\end{aligned}$$

where the four constants a_i $(i = 1, 2, 3)$ and b are given by:

$$a_i = -2\mu\varepsilon_i^o \qquad (5.2.5)$$

and

$$b = \mu(\varepsilon_1^{o2} + \varepsilon_2^{o2} + \varepsilon_3^{o2}) + d_o. \qquad (5.2.6)$$

The presence of d_o in the last equation indicates that b is not entirely determined by ε°. For this reason, b will be treated as a constant of its own, independent of ε° and hence of the other three constants a_i appearing in eq. (5.2.4).

Before proceeding further, it may be useful to recall here that according to the analysis of Sections 4.6 and 4.7, tensor **S** appearing in eq. (5.2.4) is the rotation tensor to be applied to the triad of principal axes of $\boldsymbol{\varepsilon}^\circ$ to superimpose them onto the triad of principal axes of tensor $\boldsymbol{\varepsilon}$. The latter is the actual strain of the material, measured from its macroscopic stress-free – but usually plastically deformed and microscopically distorted – state. As remarked at the end of Sect. 2.3, it may be reasonable to assume that tensor $\boldsymbol{\varepsilon}^\circ$ is deviatoric.

5.3 ALTERNATIVE EXPRESSIONS OF $\overline{\psi}_\mu$

In principal notation, tensor $\boldsymbol{\varepsilon}$ is expressed by $(\varepsilon_1, \varepsilon_2, \varepsilon_3, \mathbf{Q})$. Rotation **Q**, appearing here, represents the rotation to be imposed to the co-ordinate system (x_1, x_2, x_3) to superimpose them to the principal axes of $\boldsymbol{\varepsilon}$ [cf. Sect. 4.1]. Tensor **Q**, therefore, has a more immediate physical meaning than tensor **S**, which also depends on $\boldsymbol{\varepsilon}^\circ$ [cf. eq. (4.7.6) and Fig 4.6.2]. For this reason, we may prefer to express $\overline{\psi}_\mu$ as a function of **Q** rather than **S**, although, of course, eq. (5.2.4) is quite correct. This can easily be done by introducing eq. (4.7.6) into eq. (5.2.4), thus obtaining:

$$\overline{\psi}_{\mu} = \mu(\varepsilon_1{}^2 + \varepsilon_2{}^2 + \varepsilon_3{}^2) + a_i\, Q_{jk}\, R^{o}_{ik}\, \varepsilon_j + b. \qquad (5.3.1)$$

The following comments concerning the above equation are in order.

5.3.1 Anisotropy

The appearance of **Q** in eq. (5.3.1) shows that $\overline{\psi}_{\mu}$ depends on the principal directions of $\boldsymbol{\varepsilon}$, quite apart from the principal values of the latter. As observed in Sect. 4.3, this is the hallmark of anisotropy. Therefore, although $\overline{\psi}_{\mu}$ is isotropic in $\boldsymbol{\varepsilon}^*$ [cf. eq. (5.2.1)], it is not in general isotropic in $\boldsymbol{\varepsilon}$. It will be shown, this fact entails that the yield surface of a cold worked von Mises material is not in general an isotropic function of $\boldsymbol{\varepsilon}$ – which is not an unwelcome conclusion, since it is consistent with experimental evidence.

5.3.2 Influence of plastic deformation

Since elastic deformation does not produce any change in $\boldsymbol{\varepsilon}^{\circ}$, it leaves a_i and b unaltered; cf. eqs (5.2.5) and (5.2.6). On the contrary, plastic deformation will in general produce a change in the defect content of the material and thus in the value of $\boldsymbol{\varepsilon}^{\circ}$. This will in turn modify the values of ε°_i and $\mathbf{R}^{\circ}$, since $\boldsymbol{\varepsilon}^{\circ} \equiv \{\varepsilon^{\circ}_1, \varepsilon^{\circ}_2, \varepsilon^{\circ}_3, \mathbf{R}^{\circ}\}$. A glance at eqs (5.3.1) and (5.2.5)–(5.2.6) suffices to show that plastic deformation can affect $\overline{\psi}_{\mu}$ both through the values of a_i and b, and through the value of $\mathbf{R}^{\circ}$. Equation (5.3.1) is general enough to account for both the effects.

5.3.3 Influence of rigid-body rotation

A further comment concerns the influence of rigid-body rotation of the element of material to which eq. (5.3.1) is meant to apply. Since tensor $\boldsymbol{\varepsilon}^{\circ}$ is embedded in the material, this rotation will make its principal directions rotate with respect to the reference axes. If the rigid-body rotation of the element is represented by tensor **R**, then the triad of principal directions of $\boldsymbol{\varepsilon}^{\circ}$ will rotate from its original orientation determined by $\mathbf{R}^{\circ}$ to the new orientation determined by $\mathbf{R}\mathbf{R}^{\circ}$. Accordingly, if allowance for rigid rotation is to be made, the term R°_{ik} appearing in eq. (5.3.1) should simply be changed into $R_{is}R^{\circ}_{sk}$. Consequently, the expression of $\overline{\psi}_{\mu}$ should read:

$$\overline{\psi}_{\mu} = \mu(\varepsilon_1{}^2 + \varepsilon_2{}^2 + \varepsilon_3{}^2) + a_i\, Q_{jk}\, R_{is}R^{\circ}_{sk}\, \varepsilon_j + b. \qquad (5.3.2)$$

It should also be observed that the considered rigid-body rotation of the element might result either from the rotation of the whole body to which the element belongs or from a non-uniform deformation process. In general, the deformation process may not be infinitesimal in the usual sense of Continuum Mechanics. This would require a more appropriate definition of strain than the small strain definition thus far adopted, which in turn would bring us in the realm of finite deformations – a matter that we shall not pursue any further.

On the other hand, the dependence of $\overline{\psi}_\mu$ on the rigid-body rotation of the material is not surprising, as, in general, this quantity is not an isotropic function of strain.

5.3.4 Influence of rotation of the reference axes

In order to see how a rotation of the axes of the reference system (x_1, x_2, x_3) will modify the expression of $\overline{\psi}_\mu$, let us denote the orthogonal tensor defining this rotation by $\mathbf{P}$. Of course, a rotation of the reference axes will change the angles that these axes form with the principal directions of $\boldsymbol{\varepsilon}^\circ$. If $\bar{\mathbf{R}}^\circ$ denotes the value of tensor $\mathbf{R}^\circ$ after a rotation $\mathbf{P}$ of the same axes, we can clearly write:

$$\mathbf{R}^\circ = \mathbf{P}\,\bar{\mathbf{R}}^\circ, \tag{5.3.3}$$

since a rotation of the reference axes leaves the principal directions of $\boldsymbol{\varepsilon}^\circ$ unaltered. From this equation, we get:

$$\bar{\mathbf{R}}^\circ = \mathbf{P}^\mathsf{T}\mathbf{R}^\circ. \tag{5.3.4}$$

The expression of $\overline{\psi}_\mu$ in the rotated reference system can be obtained from eq. (5.3.1) after substituting $\bar{\mathbf{R}}^\circ$ for $\mathbf{R}^\circ$. By using eq. (5.3.4), we thus get:

$$\overline{\psi}_\mu = \mu({\varepsilon_1}^2 + {\varepsilon_2}^2 + {\varepsilon_3}^2) + a_i\, Q_{jk}\, P_{si}\, R^\circ_{sk} \varepsilon_j + b, \tag{5.3.5}$$

where, of course, tensor components Q_{jk} refer to the angles of the principal directions of $\boldsymbol{\varepsilon}$ with respect to the rotated reference axes.

On the other hand, if both rigid-body and reference axis rotations are simultaneously considered, the expression of $\overline{\psi}_\mu$ becomes:

$$\overline{\psi}_\mu = \mu({\varepsilon_1}^2 + {\varepsilon_2}^2 + {\varepsilon_3}^2) + a_i\, Q_{jk} R_{ir} P_{sr} R^\circ_{sk} \varepsilon_j + b. \tag{5.3.6}$$

This somehow clumsy equation follows on from eq. (5.3.2) once we substitute $\bar{\mathbf{R}}^\circ$ for $\mathbf{R}^\circ$ and exploit eq. (5.3.4).

Of course, if the co-ordinate axes of the reference system (x_1, x_2, x_3) are coaxial with the principal directions of $\boldsymbol{\varepsilon}°$, we have $\mathbf{R}° = \mathbf{P} = \mathbf{1}$. Moreover $\mathbf{R} = \mathbf{1}$, if rigid-body rotations are excluded. With these provisions, eq. (5.3.6) simplifies to:

$$\overline{\psi}_\mu = \mu(\varepsilon_1{}^2 + \varepsilon_2{}^2 + \varepsilon_3{}^2) + a_i\, Q_{ji}\, \varepsilon_j + b. \qquad (5.3.7)$$

5.4 DEVIATORIC COMPONENT OF THE MATRIX STRAIN ENERGY

As stated by eq. (5.2.2), the deviatoric component $\overline{\psi}_{dev}$ of the elastic energy of the matrix coincides with the analogous component of $\overline{\psi}_\mu$. The latter can readily be calculated from any of the expressions of the previous section, once we substitute the strain deviator $\mathbf{e}$ [see eq. (A.5)] for $\boldsymbol{\varepsilon}$. (Remember that, as already observed, constant b represents an amount of deviatoric elastic energy.) By referring for simplicity's sake to eq. (5.3.1), we can thus obtain:

$$\overline{\psi}_{dev} = \overline{\psi}_{dev}(e_1, e_2, e_3, \mathbf{Q}) = \mu(e_1{}^2 + e_2{}^2 + e_3{}^2) + a_i\, Q_{jk}\, R^{\circ}_{ik}\, e_j + b. \quad (5.4.1)$$

Here e_i are the principal values of $\mathbf{e}$, while the quantities a_i are defined by eq. (5.2.5). By remembering eqs (A.6) and (A.7) and by using a similar algebra to that followed in writing eq. (A.27), we can write eq. (5.4.1) in terms of $\boldsymbol{\varepsilon}$, rather than $\mathbf{e}$, as follows:

$$\overline{\psi}_{dev} = \overline{\psi}_{dev}(\varepsilon_1, \varepsilon_2, \varepsilon_3, \mathbf{Q}) = \frac{1}{3}\mu[(\varepsilon_1 - \varepsilon_2)^2 + (\varepsilon_1 - \varepsilon_3)^2 + (\varepsilon_2 - \varepsilon_3)^2]$$
$$+ a_i\, Q_{jk}\, R^{\circ}_{ik}(\varepsilon_j - \bar{\varepsilon}\delta_j) + b. \qquad (5.4.2)$$

Here $\bar{\varepsilon}$ is the mean strain, as defined by eq. (A.7), whereas δ_j is defined thus: $\delta_j = 1$ for $j = 1, 2, 3$.

From eqs (5.3.2), (5.3.5) and (5.3.6) further alternative expressions of $\overline{\psi}_{dev}$ can immediately be obtained, which take into account the rigid-body rotation of the material or the rotation of the reference axes or both. In particular, from eq. (5.3.6) we get:

$$\overline{\psi}_{dev} = \mu(e_1{}^2 + e_2{}^2 + e_3{}^2) + a_i\, Q_{jk} R_{ir} P_{sr} R^{\circ}_{sk}\, e_j + b, \qquad (5.4.3)$$

which takes into account both rigid-body rotation and rotation of the reference axes.

5.5 TAKING TOTAL STRAIN AS AN ELASTIC STRAIN MEASURE

In the presence of plastic strain, total strain is clearly different from elastic strain. For small strains, plastic, elastic and total strains are related to each other by eq. (1.2.2). Let us consider, however, an element of material in a stress-free but, in general, plastically deformed state. If deformation is measured by taking such a stress-free state as the reference for strain, elastic and total strain will coincide as long as no further plastic deformation takes place. With this proviso, all the formulae of the present chapter that involve total strain (i.e. tensor $\boldsymbol{\varepsilon}$) apply irrespective of the previous plastic deformation of the element. This has already been remarked immediately after eq. (5.1.1). The same remark is repeated here in order to avoid possible misunderstanding in the forthcoming analysis.

In the chapters that follow, we shall repeatedly adopt the expedient of referring the strain tensor of the material to its plastically deformed stress-free state, rather than to any other initial state that has been fixed once and for all. Since no further plastic strain occurs until the material is kept in its actual elastic range, total strain and elastic strain will coincide in that range, provided that they are measured from the actual stress-free state of the material, no matter how large its previous plastic deformation.

Such a procedure eliminates the need to distinguish between elastic and total strain, when discussing the yield condition of a plastically deformed material. It makes for typographically more readable formulae (in that it eliminates the apex ^{e} from $\boldsymbol{\varepsilon}$) and makes sense from the physical standpoint too. It should be borne in mind, however, that the strain space description of the phenomenon that is thus obtained is in fact an elastic-strain space description. As such, it is more akin to a stress space description rather than to a true strain space description. The latter is usually made in terms of total strain from the initial reference configuration of the virgin material.

6

Subsequent Yield Surfaces of von Mises Materials

6.1 THE FAMILY OF ALL SUBSEQUENT YIELD SURFACES

The analysis thus far developed enables us to determine the explicit expression of all the possible subsequent yield surfaces that a von Mises material can exhibit. The general yield surface of a cold worked von Mises material is expressed in symbolic form by eq. (3.3.1). Let us for the moment refer to the case in which the reference axes are fixed and rigid-body rotation of the material is excluded. In this case, the expression of $\overline{\psi}_{dev}$ is given by eq. (5.4.1), which, when introduced into eq. (3.3.1), gives:

$$G\,({e_1}^2 + {e_2}^2 + {e_3}^2) + a_i Q_{jk} R^{\circ}_{ik} e_j + b = \kappa^2, \qquad (6.1.1)$$

since $\mu \equiv G$. If we make use of eq. (5.4.2) instead of eq. (5.4.1) and remember eq. (A.6), we can alternatively express eq. (3.3.1) in terms of the principal values of $\boldsymbol{\varepsilon}$ as follows:

$$\frac{1}{3}G[(\varepsilon_1-\varepsilon_2)^2+(\varepsilon_1-\varepsilon_3)^2+(\varepsilon_2-\varepsilon_3)^2]+a_i Q_{jk} R^{\circ}_{ik}(\varepsilon_j-\overline{\varepsilon}\delta_j)+b = \kappa^2. \quad (6.1.2)$$

Equation (6.1.1) or (6.1.2) represents the family of all the possible yield surfaces of a von Mises material as a function of the principal values of **e** and

$\boldsymbol{\varepsilon}$, respectively. (Remember that in this chapter, as in the previous one, we are consistently using $\boldsymbol{\varepsilon}$ and $\mathbf{e}$ for $\boldsymbol{\varepsilon}^e$ and $\mathbf{e}^e$, since the deformations we are considering are understood to be measured from the actual stress-free state of the material and, moreover, all the processes to be considered are elastic.) Each element of the family is characterized by different values of R°_{ik}, a_i and b. Since these parameters depend on plastic deformation, each element corresponds to a different state of plastic deformation of the material.

If we also want to account for the effect of rigid-body rotation and that of reference axis rotation, eq. (5.4.3) rather than eq. (5.4.1) should be introduced into eq. (3.3.1). The family of all the subsequent yield surfaces would then become:

$$G(e_1{}^2 + e_2{}^2 + e_3{}^2) + a_i Q_{jk} R_{ir} P_{sr} R^\circ_{sk} e_j + b = \kappa^2. \tag{6.1.3}$$

In what follows, however, we shall usually refer to eq. (6.1.1) rather than the more general eq. (6.1.3). This will keep the forthcoming formulae simpler, without concealing the gist of the arguments. The effect of rigid-body rotation and that of reference axis rotation can almost immediately be included in the analysis by referring to eq. (6.1.3) instead of eq. (6.1.1).

It should be noted that the presence of $\mathbf{Q}$ in the above equations indicates that they depend on the principal directions of strain (remember that these directions are defined by $\mathbf{Q}$, according to the principal representation $\{\varepsilon_i, \mathbf{Q}\}$ of $\boldsymbol{\varepsilon}$). Thus, the subsequent yield surfaces of a von Mises material are, in general, both anisotropic functions of $\boldsymbol{\varepsilon}$ and plastic strain dependent.

Sometimes, we may be more interested in enquiring about the variety of shapes that the subsequent yield surfaces of a von Mises material can assume, rather than in finding out which surface should be associated to which plastic deformation process. In this case it may be helpful to observe that every possible yield surface of the material is covered by the following equation:

$$G(e_1{}^2 + e_2{}^2 + e_3{}^2) + a_i S_{ji} e_j + b = \kappa^2, \tag{6.1.4}$$

where the quantities S_{ij} stand for the components of any orthogonal tensor $\mathbf{S}$. Equation (6.1.4) reduces to eq. (6.1.1) if tensor $\mathbf{S}$ is taken as resulting from the composition of the two subsequent rotations $\mathbf{R}^\circ$ and $\mathbf{Q}$, according to eq. (4.7.6). Alternatively, if the effects of rigid-body rotation and reference axis rotation are taken into account, tensor $\mathbf{S}$ should be considered as resulting from the following sequence of rotation tensors:

$$\mathbf{S} = \mathbf{Q}\,\mathbf{R}^{\circ\mathrm{T}}\,\mathbf{P}\,\mathbf{R}^{\mathrm{T}}, \tag{6.1.5}$$

which makes eq. (6.1.4) coincide with eq. (6.1.3). It should be stressed, however, that eq. (6.1.4) is neither more nor less general than eq. (6.1.1) or (6.1.3). It is simply less specific in that it affords no details about the way in which rotation **S** is structured. Since **Q**, **R**, **P** and $\mathbf{R}^\circ$ can be any rotation whatsoever, the variety of yield surfaces covered by eq. (6.1.4) is exactly the same as that covered by eq. (6.1.1) or (6.1.3).

For virgin materials $\boldsymbol{\varepsilon}^\circ$ vanishes, which in view of eqs (5.2.5) and (5.2.6) makes $a_i = b = 0$. In this case eqs (6.1.1)–(6.1.4) become isotropic and coincide with the well-known expressions of yield surface of a virgin von Mises material (cf. eqs (3.1.4) and (3.1.5)). Notice that $\mathbf{R}^\circ \equiv \mathbf{1}$ if $\boldsymbol{\varepsilon}^\circ = \mathbf{0}$, since each direction is principal for a null tensor.

6.2 Q CROSS-SECTIONS OF THE SUBSEQUENT YIELD SURFACES

Further insight into the way in which the yield surface depends on plastic strain can be gained by analysing the structure of eqs (5.4.1) and (5.4.2), which enter the left-hand side of eqs (6.1.1) and (6.1.2), respectively. In order to do this, we shall arrange the whole set of strains into subsets containing all the strains possessing the same principal directions. The strains in each subset will thus share the same value of **Q**.

Let us then consider an element of material in a stress-free state after some plastic deformation, and let us enquire about its actual elastic range. The stress-free state of the element will be taken as the reference state for strain. As already observed, this will make total strain coincide with the elastic strain, as long as the actual elastic range of the material is not exceeded. Coefficients a_i and b will remain constant within the elastic range because they are defined by eqs (5.2.5) and (5.2.6), and purely elastic processes do not affect the value of $\boldsymbol{\varepsilon}^\circ$. Tensor $\mathbf{R}^\circ$ will not change either, since it represents the directions of the principal axes of $\boldsymbol{\varepsilon}^\circ$ with respect to a fixed reference system. Since **Q** is constant in each of the strain subsets defined above, the elastic range relevant to the states of strain belonging to any one of these subsets will be expressed by the following quadratic form:

$$G({e_1}^2 + {e_2}^2 + {e_3}^2) + \bar{a}_j e_j + b = \kappa^2. \tag{6.2.1}$$

Constants $\bar{a}_j$ appearing in the above equation are given by:

$$\bar{a}_j = a_i\, Q_{jk}\, R^\circ_{ik}. \tag{6.2.2}$$

Equation (6.2.1) represents a cross-section at constant **Q** of the yield surface of the material. It will be called **Q** cross-section of the yield surface. According to eq. (6.2.2), this cross-section depends on the previous plastic deformation since constants $\bar{a}_j$ so do through a_i and R°. It will also depend on the principal directions of the considered subset of strains since constants $\bar{a}_j$ depend also on **Q**.

The **Q** cross-sections of the yield surface can be given an interesting geometric representation in space $(\psi_{dev}, e_1, e_2, e_3)$. This space is more convenient than the space $(\psi_{dev}, \varepsilon_1, \varepsilon_2, \varepsilon_3)$, since $\overline{\psi}_{dev}$ depends on strain through strain deviator **e**. A simple glance at eq. (5.4.1) shows that the deviatoric elastic energy at constant **Q** of the cold worked material differs from that of the virgin material (eq. (A.27)) by a linear term in e_i. Geometrically speaking, this means that in the space $(\psi_{dev}, e_1, e_2, e_3)$ surface $\overline{\psi}_{dev} = \overline{\psi}_{dev}(e_1, e_2, e_3, \mathbf{Q})$ results from a rigid-body translation of the deviatoric energy surface ψ°_{dev} of the virgin material. The latter surface is a (hyper-) paraboloid of rotation about the ψ-axis, as apparent from its analytical expression (A.27). Thus, all the **Q** cross-sections of the deviatoric energy of the matrix of a cold worked material are in fact (hyper-)paraboloids, similar to the virgin one and obtained from the latter by appropriate rigid-body translations.

The translation will depend on the plastic deformation of the material through the quantities a_i, b and **R**°. It will also depend on the direction of strain, since **Q** enters the definition of constants $\bar{a}_j$, according to eq. (6.2.2). (Remember that the deviatoric energy function of the matrix of a cold worked von Mises material is not in general isotropic, although that of the virgin material is.)

It should also be observed that these rigid-body translations are completely defined by the displacement of the paraboloid axis. This displacement coincides with the displacement of the paraboloid point of minimum, since the latter lies in the paraboloid axis. For virgin materials, the point of minimum coincides with the origin of coordinate system $(\psi_{dev}, e_1, e_2, e_3)$. This is apparent from eq. $(A.27)_5$, since we assume that the strain energy of the virgin material vanishes at the stress-free state, which makes $c_o = 0$. On the other hand, for cold worked materials, the coordinates of the paraboloid minimum depend on plastic deformation. They can be calculated by setting the derivatives of eq. (5.4.1) with respect to e_i equal to zero. By making use of eqs (5.2.5) and (4.7.6), we thus get:

$$e_i|_{min} = Q_{ir}\, R^{\circ}_{kr}\, \varepsilon^{\circ}_k = S_{ik}\varepsilon^{\circ}_k. \tag{6.2.3}$$

A comparison of this equation with eq. (4.6.5) shows that the coordinates of minimum of the considered paraboloids are defined by the same vector $\mathbf{p}^\circ$ that we introduced in Sect. 4.6. In the present case, we only have to identify the principal components of tensor $\boldsymbol{\varepsilon}^\circ$ appearing in eq. (4.6.5) with those of the permanent elastic strain $\boldsymbol{\varepsilon}^\circ$ as defined in Sect. 5.1 (it should be borne in mind that, as observed in Sect. 2.3, tensor $\boldsymbol{\varepsilon}^\circ$ is supposed to be deviatoric).

The scenario described above is represented in Fig. 6.2.1. As usual, the figure refers to the abridged three-dimensional space (ψ_{dev}, e_1, e_2), rather than the full four-dimensional space $(\psi_{dev}, e_1, e_2, e_3)$. The deviatoric energy of the virgin material, is represented by just one single paraboloid, the axis of which coincides with the ψ_{dev}-axis of the coordinate system. It is denoted by $\overline{\psi}^\circ_{dev}$ in the figure. The paraboloid is one and the same, no matter the value of $\mathbf{Q}$. This is exactly how it should be, owing to the assumed isotropy of the virgin material. Its analytical expression is:

$$\overline{\psi}^\circ_{dev} = G({e_1}^2 + {e_2}^2 + {e_3}^2), \tag{6.2.4}$$

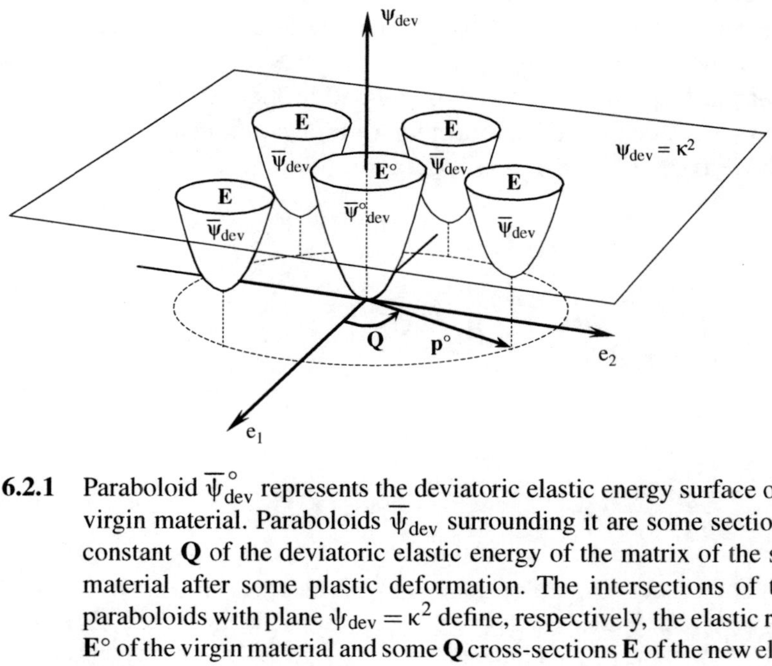

Fig. 6.2.1 Paraboloid $\overline{\psi}^\circ_{dev}$ represents the deviatoric elastic energy surface of the virgin material. Paraboloids $\overline{\psi}_{dev}$ surrounding it are some sections at constant $\mathbf{Q}$ of the deviatoric elastic energy of the matrix of the same material after some plastic deformation. The intersections of these paraboloids with plane $\psi_{dev} = \kappa^2$ define, respectively, the elastic range $\mathbf{E}^\circ$ of the virgin material and some $\mathbf{Q}$ cross-sections $\mathbf{E}$ of the new elastic range.

as can easily be obtained from eq. (5.1.1) by taking the deviatoric part of $\overline{\psi}^{\circ}(\boldsymbol{\varepsilon})$.

On the other hand, in the same figure the paraboloids denoted by $\overline{\psi}_{\text{dev}}$ refer to the deviatoric energy of the matrix of the cold worked material, as defined by eq. (5.4.1). All of them are supposed to be relevant to the same state of plastic deformation of the material. Each paraboloid corresponds to a different value of **Q**, though. Any one of them can be made to coincide with the others or, for that matter, with the virgin one by a rigid-body translation.

From eq. (6.1.1) it follows that the intersections of plane $\psi_{\text{dev}} = \kappa^2$ with the above paraboloids define the elastic domain of the material. In particular, the intersection of the above plane with the paraboloid $\overline{\psi}^{\circ}_{\text{dev}}$ delimits the elastic domain of the virgin material. It is denoted by $\mathbf{E}^{\circ}$ in Fig. 6.2.1. The analogous intersections with the $\overline{\psi}_{\text{dev}}$ paraboloids represent a few **Q** cross-sections of the yield surface of the cold worked material. In the figure, the regions within these cross-sections are all denoted by the same symbol **E**. Each region refers to a different value of **Q** and, therefore, represents a different **Q** cross-section of the elastic range **E** of the cold worked material.

We have already observed that the plastic strain of the material affects the value of $\boldsymbol{\varepsilon}^{\circ}$ and hence the values of constants a_i and b. In the geometric representation of Fig. 6.2.1, this means that as the plastic deformation of the material varies, the $\overline{\psi}_{\text{dev}}$ paraboloids change both in their distance from the origin (represented by vector $\mathbf{p}^{\circ}$) and in the value of their minimum. Figure 6.2.2 depicts three different **Q** cross-sections of $\overline{\psi}_{\text{dev}}$ (variously denoted as $\mathbf{E}'$, $\mathbf{E}''$ and $\mathbf{E}'''$ in the figure) relevant to three different plastic

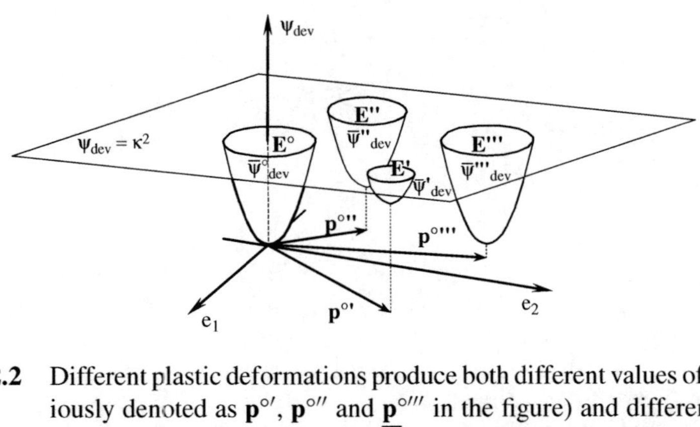

Fig. 6.2.2 Different plastic deformations produce both different values of $\mathbf{p}^{\circ}$ (variously denoted as $\mathbf{p}^{\circ\prime}$, $\mathbf{p}^{\circ\prime\prime}$ and $\mathbf{p}^{\circ\prime\prime\prime}$ in the figure) and different values of the point of minimum of the $\overline{\psi}_{\text{dev}}$ paraboloids. This generates a sort of kinematic and isotropic hardening in the **Q** cross-sections of the relevant elastic ranges of the material.

deformation processes of the material. It shows that a change of $\mathbf{p}^{\circ}$ (i.e. a change of plastic deformation) produces a rigid-body translation of the elastic domain $\mathbf{E}$ in the $(\overline{\psi}_{\text{dev}} = \text{const})$-plane, while a change of the minimum value of $\overline{\psi}_{\text{dev}}$ makes the same domain grow or shrink uniformly.

Such a dependence of the elastic range on plastic deformation, however, should not be taken as indicating the presence of what is usually referred to as kinematic hardening and isotropic hardening, respectively. Far from this. These two kinds of hardening are defined in stress space, while the changes in the elastic regions described above refer to their representation in quite a different space, be it space $(\psi_{\text{dev}}, e_1, e_2, e_3, \mathbf{Q})$ or, if we prefer, its subspace at constant ψ_{dev}, say $(e_1, e_2, e_3, \mathbf{Q})$, or equivalently, $(e_1, e_2, e_3, \phi, \theta, \psi)$. Here the quantities φ, θ, and ψ are the three Euler's angles, which can be used to define the orthogonal tensor $\mathbf{Q}$ completely, as indicated by eq. (B.20) of Appendix B. Any translation or growth of the elastic domains in these spaces maps distortedly in the stress space, so that it does not correspond to a similar translation or growth of the corresponding elastic domains in stress space. The reason is that the stress space is not linearly related to the space $(e_1, e_2, e_3, \mathbf{Q})$ or $(e_1, e_2, e_3, \phi, \theta, \psi)$.

To be convinced of the latter point, let us consider all the states of stress possessing a certain given triad of principal directions. In space $(e_1, e_2, e_3, \phi, \theta, \psi)$ these states of stress correspond to a region at constant values of coordinate variables ϕ, θ and ψ. (Remember that, since $\mathbf{e}$ is understood here to denote elastic strains, tensors $\mathbf{e}$ and $\boldsymbol{\sigma}$ are coaxial.) However, no matter how we choose the coordinate axes, the same states of stress cannot be covered by a region in stress space $(\sigma_{11}, \sigma_{22}, \sigma_{33}, \sigma_{12}, \sigma_{23}, \sigma_{13})$ where three stress components remain constant. This is a consequence of the fact that the states of stress that possess a given triad of principal directions are not linearly related to each other and, therefore, their components cannot all lie on the same plane.

Take the particular states of stress defined by eq. (1.3.2), for instance. All these stresses belong to plane (σ, τ) in the stress space. However, angle θ_3 defining the principal directions of these states of stress can be anything we like since it depends on the ratio σ/τ. Due to the non-linear relationship (1.3.4) between σ, τ and θ_3, this plane region of the stress space cannot be mapped in a two-dimensional plane in $(e_1, e_2, e_3, \phi, \theta, \psi)$-space.

6.3 YIELD SURFACES IN PRACTICE

Though fairly simple, eq. (6.1.1) is capable of producing elastic domains in an astonishing variety of sizes and shapes when applied to determine the two-dimensional elastic regions usually considered in experimental plasticity. Such a wealth of possibilities agrees well with the vast assortment of seemingly unrelated shapes that the elastic regions of ductile metals show when tested. Usually, the outcome of the experiments is expressed in terms of the stress components that are controlled during the test. The corresponding theoretical predictions can be obtained directly from eq. (6.1.1), once it is recalled that the strain components e_i appearing there are intended to be elastic and, therefore, can immediately be expressed in terms of stress through Hooke's law. The expression of the desired limit curve or limit surface in terms of the desired stress components then becomes a matter of routine algebraic manipulation.

An instance of this derivation is presented in the next section, where the subsequent elastic regions relevant to tension/torsion tests of thin-walled tubes will be considered. The curves limiting these regions will be obtained from eq. (6.1.1) by expressing the strain variables e_i appearing there in terms of the stress components that are controlled during the test.

A point to be made here is that all information concerning the elastic limits of the material is in fact contained in eq. (6.1.1). Thus, if this equation is given, the knowledge of the limit curves or limit surfaces relevant to any particular family of processes can be dispensed with altogether. The elastic limit of the material can always be obtained directly from eq. (6.1.1), no matter which family of process is considered. Strictly speaking, therefore, there is no need to make any reference whatsoever to particular stress or strain spaces of reduced dimensions or to elastic subdomains bounded by appropriate limit curves or limit surfaces of reduced dimensions.

It should be noted, however, that in order to apply the general equation (6.1.1) we need to know the value of $\boldsymbol{\varepsilon}^\circ$ at any time during the process, since a_i, b and $\mathbf{R}^\circ$ depend upon it. The evolution rule for $\boldsymbol{\varepsilon}^\circ$ will be considered in the next chapter. It will enable us to predict how the elastic domain changes as plastic deformation proceeds and will play the role of the so-called work-hardening rule.

Although it is true that eq. (6.1.1) eliminates the need to deal with elastic domains of reduced dimensions, there are at least two good reasons why we may need to deal with reduced domains relevant to particular families of processes. The first one is that in many practical situations, the curves bounding these domains enable us to verify at a glance the extent to which the theoretical predictions fit the experimental data. The second reason is that the way in which these curves evolve with plastic deformation gives us a most effective clue as to how to establish realistic work-hardening rules for specific materials.

6.4 SUBSEQUENT LIMIT CURVES FOR STANDARD TENSION/TORSION TESTS

Apart from the elementary case of uniform pressure, a three axial state of stress cannot be controlled easily in practice. This is why most of the experiments in metal plasticity concern biaxial tests. A widely used one is the tension/torsion test of thin-walled tubes. It is performed by applying axial force N and torque T to the ends of thin-walled tubular specimens of circular cross-section (see Fig. 6.4.1a). The ensuing state of stress is biaxial and uniform throughout the specimen. It is to all purposes equivalent to that produced by applying axial stress and shear stress to a flat sheet of material, as specified in Fig. 6.4.1b. Axial stress and shear stress at the specimen cross-sections are respectively given by $\sigma = N/A$ and $\tau = T/(A\,r)$. Here, A is the cross-section area of the specimen, while r is its mean radius (the average between its inner and outer radius). If reference is made to the coordinate system (x_1, x_2, x_3) defined in Fig. 6.4.1, the components of stress tensor $\boldsymbol{\sigma}$ at any point of the specimen are the ones we already specified when writing eq. (1.3.2). In a tension/torsion test, therefore, the state of stress of the specimen is entirely controlled by stress components σ and τ or equivalently by N and T. Whether performed on thin-walled tubes or flat bars, this test is often referred to as N/T test.

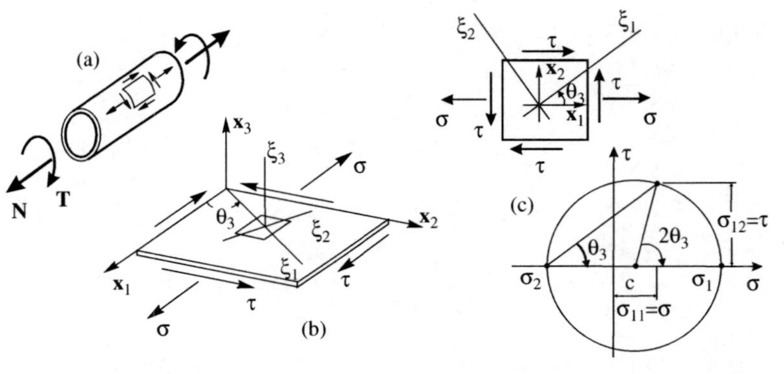

Fig. 6.4.1 (a) Tension/torsion of a thin-walled tube. (b) Equivalent plane state of stress of a plane sheet of material and placement of principal axes ξ_1, ξ_2 and ξ_3. (c) Stress components in a specimen element and relevant Mohr's circle.

As already recalled in Sect. 1.3, during the test the principal stresses of the specimen are given by:

$$\sigma_1 = \frac{\sigma}{2} + \sqrt{\sigma^2/4 + \tau^2}, \quad \sigma_2 = \frac{\sigma}{2} - \sqrt{\sigma^2/4 + \tau^2}, \quad \sigma_3 = 0. \qquad (6.4.1)$$

The principal axes relevant to these principal values will be denoted by ξ_1, ξ_2 and ξ_3, respectively. In the present case, axis ξ_3 will be parallel to the coordinate axis x_3, while the other two will lie on the (x_1, x_2)-plane. The placement of all these axes is therefore determined by the angle θ_3 that the x_1-axis forms with the ξ_1-axis. More precisely, this angle will be defined as the rotation about the x_3-axis that makes axes x_1 and x_2 superimpose onto axes ξ_1 and ξ_2, respectively. By assuming that a positive value of θ_3 represents an anticlockwise rotation when viewed from the positive side of the x_3-axis, angle θ_3 is given by eq. (1.3.4), which we note again here for convenience:

$$\theta_3 = \frac{1}{2} \operatorname{atan} \frac{2\tau}{\sigma}. \qquad (6.4.2)$$

Further references on these well-known formulae can be found in any standard textbook on Continuum Mechanics (see [6, Sect. 3.6]).

In writing eqs (6.4.1), we adhered to the standard convention according to which the principal values of $\boldsymbol{\sigma}$ are labelled in such a way that $\sigma_1 \geq \sigma_2$. As the values of the applied forces N and T are changed, axis ξ_1 will rotate to make the angle θ_3 with axis x_1 defined by eq. (6.4.2). By referring to Fig. 6.4.1c and to Mohr's circle represented there, it is not difficult to verify that, depending on the value of the applied stress, the values of θ_3 range within the following limits:

$$-\frac{\pi}{4} \leq \theta_3 \leq \frac{\pi}{4} \quad \text{for } \sigma \geq 0, \qquad (6.4.3)$$

$$\frac{\pi}{4} \leq \theta_3 \leq \frac{\pi}{2} \quad \text{for } \sigma \leq 0 \text{ and } \tau \geq 0, \qquad (6.4.4)$$

$$-\frac{\pi}{2} \leq \theta_3 \leq -\frac{\pi}{4} \quad \text{for } \sigma \leq 0 \text{ and } \tau \leq 0. \qquad (6.4.5)$$

We can now proceed to calculate the principal values attained by elastic strain deviator $\mathbf{e}$ during the test. From eqs (6.4.3)–(6.4.5), (6.4.1), (1.2.16) and (1.2.17) the principal components of stress deviator $\mathbf{s}$ can easily be expressed as functions of σ and τ. From eq. (A.8), Hooke's law (A.25) and eq. (A.14), the principal values of $\mathbf{e}$ can then be expressed as functions of the same variables σ and τ as follows:

$$e_1 = \frac{1+\nu}{6\,E}(\sigma + 3\sqrt{\sigma^2 + 4\tau^2}), \tag{6.4.6}$$

$$e_2 = \frac{1+\nu}{6\,E}(\sigma - 3\sqrt{\sigma^2 + 4\tau^2}), \tag{6.4.7}$$

$$e_3 = -2\frac{1+\nu}{6\,E}\sigma. \tag{6.4.8}$$

(For the record, it should be mentioned here that eqs (6.4.3)–(6.4.8) correct similar equations that were mistyped in [7] and [8].)

In order to apply eq. (6.1.1), we also need to express the components of tensor **Q** in terms of the controlling variables σ and τ. Remember that tensor **Q** represents the rotation to be imposed to the axes of the reference triad to superimpose them onto the principal triad of **e**. The latter triad coincides with the principal triad of elastic strain **ε**, since a symmetric second order tensor and its deviatoric part share the same principal directions. On the other hand, tensor **ε** is coaxial with **σ**, since von Mises materials are isotropic in the elastic range. It follows that the rotation defined by **Q** coincides with the rotation to be applied to the axes of coordinate system (x_1, x_2, x_3) to superimpose them onto the principal axes of **σ**. From eq. (6.4.2) and from the meaning of the components of **Q**, as recalled in Sect. 4.1, it then follows that:

$$\mathbf{Q} \equiv \begin{bmatrix} \cos\theta_3 & \sin\theta_3 & 0 \\ -\sin\theta_3 & \cos\theta_3 & 0 \\ 0 & 0 & 1 \end{bmatrix}. \tag{6.4.9}$$

In order to determine the σ/τ yield curves relevant to the considered tension/torsion tests, we also have to specify the value of $\mathbf{R}^\circ$ appearing in eq. (6.1.1). This tensor defines the rotation of the principal directions of permanent elastic strain $\boldsymbol{\varepsilon}^\circ$ with respect to the reference axes. The knowledge of $\boldsymbol{\varepsilon}^\circ$ is also needed to determine, from eqs (5.2.5) and (5.2.6), the coefficients a_i and b, which also appear in eq. (6.1.1). The tensor $\boldsymbol{\varepsilon}^\circ$, however, depends on the plastic deformation history of the material and can only be determined if its evolution rule is known, which will only be available in the next chapter. For this reason, in the present section we shall content ourselves with presenting just a few samples of possible σ/τ limit curves that are predicted by eq. (6.1.1). To do so, we shall assign some particular values to $\boldsymbol{\varepsilon}^\circ$, assuming that, for each of these values, the body suffered the appropriate elastic-plastic deformation history leading to that value. Our purpose here is simply to check whether the outcome of eq. (6.1.1) is rich enough to represent the variety of elastic domains that a ductile metal can exhibit when subjected to tension/torsion tests.

With this aim in mind, we shall consider the particular case in which $\mathbf{R}^\circ$ represents a rotation about the x_3-axis. We shall accordingly denote by β the rotation angle about the x_3-axis that makes the coordinate system (x_1, x_2, x_3) superimpose onto the principal triad of $\boldsymbol{\varepsilon}^\circ$. In the coordinate system (x_1, x_2, x_3), the components of $\mathbf{R}^\circ$ will therefore be given by:

$$\mathbf{R}^\circ \equiv \begin{bmatrix} \cos\beta & \sin\beta & 0 \\ -\sin\beta & \cos\beta & 0 \\ 0 & 0 & 1 \end{bmatrix}, \tag{6.4.10}$$

which is analogous to eq. (6.4.9) expressing a similar rotation. Assumption (6.4.10) is not unrealistic if the test starts from the virgin material and any subsequent plastic deformation is produced in the specimen by the considered N/T processes. In this case, the state of stress of the specimen is consistently plane and the principal direction of zero stress coincides with the x_3-axis throughout the deformation process. Elementary symmetry considerations suggest then that the x_3-axis should be a principal axis for $\boldsymbol{\varepsilon}^\circ$ too , which implies that $\mathbf{R}^\circ$ should have the form (6.4.10). (The evolution law to be introduced in the next chapter will actually be consistent with this.)

Be it as it may, from eqs (6.4.10) and (6.4.9) we get:

$$\mathbf{R}^\circ\mathbf{Q}^\mathsf{T} \equiv \begin{bmatrix} \cos(\theta_3-\beta) & -\sin(\theta_3-\beta) & 0 \\ \sin(\theta_3-\beta) & \cos(\theta_3-\beta) & 0 \\ 0 & 0 & 1 \end{bmatrix}. \tag{6.4.11}$$

By introducing this equation and eqs (6.4.6)–(6.4.8) into eq. (6.1.1), we finally get:

$$\begin{aligned} &2\sigma^2 + 6\tau^2 + [(a_1+a_2)\cos(\theta_3-\beta) - (a_1-a_2)\sin(\theta_3-\beta) - 2a_3]\,\sigma \\ &\quad +3\,[(a_1-a_2)\cos(\theta_3-\beta) + (a_1+a_2)\sin(\theta_3-\beta)]\sqrt{\sigma^2+4\tau^2} \\ &\quad = \frac{6\,E}{\nu+1}(\kappa^2 - b), \end{aligned} \tag{6.4.12}$$

where θ_3 is determined by σ and τ according to eq. (6.4.2).

Equation (6.4.12) represents all the possible σ/τ limit curves that an initially virgin specimen of von Mises material can ever exhibit in a tension/torsion test. The plastic deformation process suffered by the specimen during the test determines the values of parameters a_i, b and β, and thus the particular limit curve it will exhibit at the end of the process. A selection of particular limit curves resulting from different choices of these parameters is presented in Fig. 6.4.2. A cursory glance at the available experimental data (cf. e.g. [2, pt. VI]) suffices to show that eq. (6.4.12) is likely to be flexible enough to reproduce to the most crucial detail the host of different and seemingly unrelated limit curves that appear in the vast literature on the subject.

It should be mentioned that for $\sigma < 0$, eq. (6.4.12) produces the mirror image with respect to σ-axis of the correct σ/τ yield curve. This shortcoming is a consequence of the fact that, as σ changes from positive to negative values, the principal stress directions 1 and 2 suffer a sudden $\pi/2$ rad rotation about the third principal axis (x_3-axis). In Mohr's construction this is equivalent to rotating by an angle of $\pi/2$ rad the reference face of the element on which σ and τ are considered. According to Mohr's convention concerning the sign of τ, such a rotation produces an inversion of the positive direction to be assumed for τ. To keep track of this, one should substitute $-\tau$ for τ when applying eq. (6.4.12) for $\sigma < 0$. This is equivalent to mirroring back, with respect to σ-axis, the part of the σ/τ yield curve that is relevant to $\sigma < 0$. This simple device will suffice to eliminate the above shortcoming.

It is important to stress that eq. (6.4.2) was obtained here in a purely deductive way, starting from the yield condition of a virgin von Mises material. The wealth of shapes and sizes of the σ/τ yield curves it produces is, therefore, a direct consequence of the virgin yield condition itself. In particular, we did not introduce any hypothesis on how the subsequent elastic domain of the

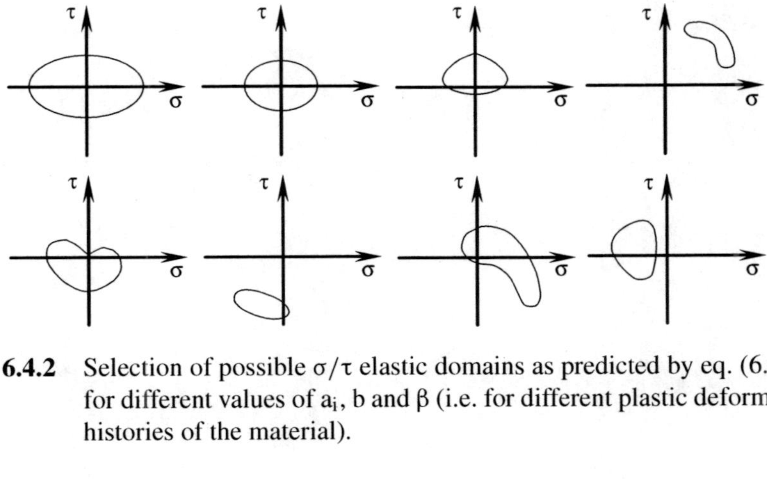

Fig. 6.4.2 Selection of possible σ/τ elastic domains as predicted by eq. (6.4.12) for different values of a_i, b and β (i.e. for different plastic deformation histories of the material).

material should change, or any assumption about its dependence on plastic deformation. It is the yield condition of the virgin material that by itself determines all the possible subsequent yield surfaces that are admissible for the material.

It should be observed, finally, that eq. (6.4.12) does not cover all the possible σ/τ limit curves that a von Mises material can ever exhibit during a tension/torsion test. If the test is performed on a specimen of a non-virgin material whose previous work-hardening process is unrestricted, tensor $\mathbf{R}^\circ$ will in general assume the form (B.20), which is much more general than the one given by eq. (6.4.11). The family of all the possible σ/τ limit curves could again be obtained by following similar lines as those leading to eq. (6.4.12). The resulting expression would however be much more complicated, though of course still easily tractable numerically. The corresponding family of yield curves would be richer than the previous one, as the component of $\mathbf{R}^\circ$ would now involve three parameters rather than just one as in the case of eq. (6.4.10). From the qualitative standpoint, however, the new family is not expected to involve elastic domains that are dramatically different in shape than those encompassed by eq. (6.4.12). The reason is that rotating the principal directions of $\boldsymbol{\varepsilon}^\circ$ makes little qualitative difference, since this tensor is geometrically represented by an ellipsoid.

7

The Work-Hardening Rule

The family of all possible yield surfaces that a work-hardened von Mises material can ever exhibit was determined in the previous chapter. We arrived at that result in a purely deductive way, starting from a few general, elementary and almost obvious hypotheses. Though gratifying, that result is not enough for most of the practical applications of the theory of plasticity. They usually aim at solving specific boundary-value problems, which invariably requires working out which of all the possible yield surfaces is to be associated to which elastic-plastic deformation process. The present chapter attempts to answer this question. It proposes a reasonably general evolution rule for $\boldsymbol{\varepsilon}^{\circ}$ to associate the appropriate subsequent yield surface to any elastic-plastic deformation process we may care to specify.

The present analysis will show how a suitable work-hardening rule can be constructed on the basis of the theory of the previous chapters and will provide the landmarks for a better interpretation of the experimental results on subsequent yielding. Of course, much work, both theoretical and experimental, remains to be done before this part of the theory can be given a definitive shape. Such work will have to take full account of the strain history dependence of the elastic-plastic response of the material.

For these reasons, the results of this chapter are not meant to be definitive or to apply to every von Mises material. For many of these materials, however, they will provide a realistic prediction of the evolution of their elastic region in the whole range of elastic-plastic deformation processes. This means, in particular, that they can be applied irrespective of whether the deformation process involves one-, two- or three-dimensional stress/strain states, no matter how far they are from the virgin state.

7.1 THE VARIABLES THAT CONTROL WORK-HARDENING IN VON MISES MATERIALS

We shall refer to work-hardening as to the change in the elastic limits of the material due to cold work. The word 'hardening' is used loosely here as cold work will either increase or reduce the elastic limit of a material, depending on which point of the yield surface is considered. Semantics apart, the yield surface of a cold worked von Mises material is fully determined by eq. (6.1.1). It depends on plastic deformation through a_i, b and $\mathbf{R}^\circ$. In a virgin material, these quantities vanish and the yield surface reduces to eq. (3.1.4).

Both a_i and $\mathbf{R}^\circ$ are determined by $\boldsymbol{\varepsilon}^\circ$. Coefficients a_i depend on the principal values of $\boldsymbol{\varepsilon}^\circ$, as specified by eq. (5.2.5), while $\mathbf{R}^\circ$ represents the rotation of the principal directions of $\boldsymbol{\varepsilon}^\circ$ with respect to the reference axes. An evolution rule for $\boldsymbol{\varepsilon}^\circ$ is, therefore, all we need to determine the values of a_i and $\mathbf{R}^\circ$ produced by an elastic-plastic deformation process.

The situation is different for constant b. This constant is defined by eq. (5.2.6) and is independent of $\boldsymbol{\varepsilon}^\circ$ due to the presence of d_o in that equation. If the yield surface is given, say from experiments, then the value of b can be determined by recalling that plastic deformation can only occur on the yield surface. This means that at any time during any elastic-plastic deformation process, the point that represents the elastic strain at which the most recent plastic deformation took place must be *on* the actual yield surface of the material. The condition that surface (6.1.1) should contain that point can be exploited to determine the value of b once $\boldsymbol{\varepsilon}^\circ$ is known. If, on the other hand, we are to predict which of all the possible yield surfaces of the material will represent the actual one after a given deformation process, then a specific rule to determine the value of b (or equivalently that of d_o) must be given in addition to the evolution rule for $\boldsymbol{\varepsilon}^\circ$.

Of course, any evolution rule for $\boldsymbol{\varepsilon}^\circ$ and d_o must meet the restrictions that come from the principles of mechanics and thermodynamics. At the present state of knowledge, however, we cannot derive such a rule by means of deductive arguments as we did when seeking the general expression of all the possible subsequent yield surfaces of a von Mises material. For want of a better approach, we shall, therefore, be content to regard the evolution rule for $\boldsymbol{\varepsilon}^\circ$ and d_o as a constitutive feature. Its expression should, accordingly, be determined from appropriate experimental tests on the particular von Mises material under consideration.

Even this is a tremendous task, though. The number of different elastic-plastic processes that a material can suffer is enormous and their outcomes are extremely varied. In such a situation, some educated guesswork has to be made about the general structure of this law before resorting to its experimental determination. Some reasonable evolution rules for $\boldsymbol{\varepsilon}^\circ$ and d_o will be proposed in the sections that follow.

7.2 AN EVOLUTION RULE FOR $\boldsymbol{\varepsilon}^\circ$

In looking for the evolution law of $\boldsymbol{\varepsilon}^\circ$, the principal notation we introduced in Sect. 4.1 for symmetric second order tensors is particularly appropriate. In that notation, tensor $\boldsymbol{\varepsilon}^\circ$ reads as $\{\varepsilon_1^\circ, \varepsilon_2^\circ, \varepsilon_3^\circ, \mathbf{R}^\circ\}$, which shows explicitly all the quantities we need to read off from $\boldsymbol{\varepsilon}^\circ$ in order to determine both a_i and $\mathbf{R}^\circ$. Accordingly, the evolution law we are looking for can be decomposed into two parts. One part will specify how the principal values of $\boldsymbol{\varepsilon}^\circ$ are affected by plastic deformation; the other will give the value of $\mathbf{R}^\circ$ (which means the principal directions of $\boldsymbol{\varepsilon}^\circ$) at any time during the process.

A simple evolution law, which is likely to be general enough to find wide application is the following. Let $\dot{\boldsymbol{\varepsilon}}^\circ$ be the increment of $\boldsymbol{\varepsilon}^\circ$ at a certain time during the deformation process. In addition, let $\dot{\mathbf{e}}$ be the increment of the deviatoric component $\mathbf{e}$ of total strain $\boldsymbol{\varepsilon}$ (as distinguished from elastic strain $\boldsymbol{\varepsilon}^e$. Notice that in this chapter, the elastic strain and the elastic strain deviator will always be explicitly denoted as $\boldsymbol{\varepsilon}^e$ and $\mathbf{e}^e$, since for processes that bring the material outside its actual elastic range the distinction between elastic and total strain is, of course, mandatory.) The proposed evolution rule assumes that the principal values $\dot{\varepsilon}_i^\circ$ of $\dot{\boldsymbol{\varepsilon}}^\circ$ are related to the analogous values $\dot{e}_i$ of $\dot{\mathbf{e}}$ as follows:

$$\dot{\varepsilon}_i^\circ = \begin{cases} \gamma(\boldsymbol{\xi})\,\dot{e}_i & \text{for elastic-plastic processes } \left(f = 0 \text{ and } \dfrac{\partial f}{\partial \sigma_{ij}}\dot{\sigma}_{ij} > 0\right) \\ 0 & \text{for purely elastic processes } \left(f < 0 \text{ or } f = 0 \text{ and } \dfrac{\partial f}{\partial \sigma_{ij}}\dot{\sigma}_{ij} \le 0\right). \end{cases} \tag{7.2.1}$$

Here $\gamma = \gamma(\boldsymbol{\xi})$ denotes a scalar function of an appropriate set of variables $\boldsymbol{\xi}$, whose number and nature have to be specified on the basis of experimental evidence. We shall refer to γ as the *reduction factor*.

In writing eq. (7.2.1), we assumed that the yield function f is expressed as a function of stress, which is the most usual choice. Of course, if f is expressed as a function of elastic strain, as has often been the case in the present book, we simply have to substitute $(\partial f/\partial \varepsilon^e_{ij})\,\dot{\varepsilon}^e_{ij}$ for $(\partial f/\partial \sigma_{ij})\dot{\sigma}_{ij}$ in the above equation.

The definition of $\dot{\boldsymbol{\varepsilon}}^\circ$ cannot be complete without assigning the directions of its principal axes. In order to do this, let α_i $(i = 1, 2, 3)$ denote the angle between the principal direction i of tensor $\dot{\mathbf{e}}$ and the homologous principal direction of tensor $\boldsymbol{\varepsilon}^\circ$. The considered evolution rule assumes that the principal direction i of tensor $\dot{\boldsymbol{\varepsilon}}^\circ$ is intermediate between the homologous principal directions of tensors $\dot{\mathbf{e}}$ and $\boldsymbol{\varepsilon}^\circ$. More precisely, it assumes that angle α_i° between direction i of $\dot{\boldsymbol{\varepsilon}}^\circ$ and direction i of $\boldsymbol{\varepsilon}^\circ$ is given by:

$$\alpha_i^\circ = r(\boldsymbol{\xi})\,\alpha_i. \qquad (7.2.2)$$

Here $r = r(\boldsymbol{\xi})$ is a scalar-valued function of $\boldsymbol{\xi}$, taking values in the interval [0,1]. Scalar r will be called *angular drag factor*, or *drag factor* for short.

The evolution rule specified by eqs (7.2.1) and (7.2.2) is represented in Fig. 7.2.1 with reference to plane tensors. The latter are portrayed as ellipsoids according to Lamé representation. Suppose that at a certain time of the process, the value of $\boldsymbol{\varepsilon}^\circ$ is known and we want to find out how this tensor changes as the plastic deformation process proceeds. The figure shows that this can be done in two steps. In the first step, we calculate the deviatoric part $\dot{\mathbf{e}}$ of the total strain increment and construct the reduced tensor $\gamma\,\dot{\mathbf{e}}$. The second step consists of calculating the angles α_i between the i-axes of $\dot{\mathbf{e}}$ and $\boldsymbol{\varepsilon}^\circ$ and then rotating tensor $\gamma\,\dot{\mathbf{e}}$, so that its principal directions form the reduced angles $\alpha_i^\circ = r\alpha_i$ with the homologous directions of $\boldsymbol{\varepsilon}^\circ$. The rotated

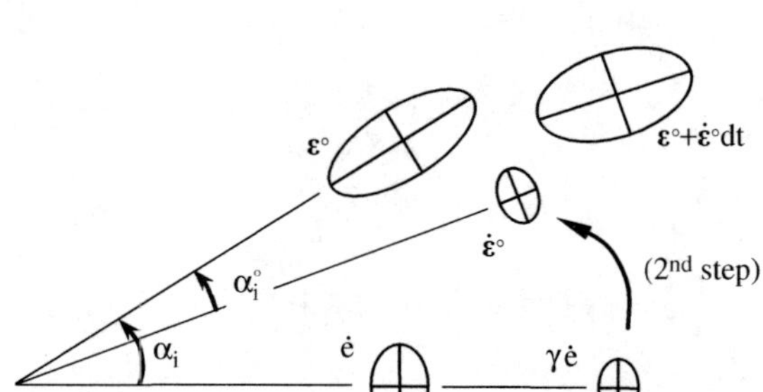

Fig. 7.2.1 Graphic representation of the two-step procedure to implement the proposed evolution law for $\boldsymbol{\varepsilon}^\circ$.

tensor thus obtained is $\dot{\boldsymbol{\varepsilon}}^\circ$. The new value of $\boldsymbol{\varepsilon}^\circ$ after the considered total strain increment will then be $\boldsymbol{\varepsilon}^\circ + \dot{\boldsymbol{\varepsilon}}^\circ\,dt$.

It should not escape the reader's attention that the above rule requires that we should know which index $i \in \{1, 2, 3\}$ is assigned to which principal direction of tensors $\dot{\mathbf{e}}$ and $\boldsymbol{\varepsilon}^\circ$. As observed in Sect. 6.4, the usual convention is to order principal values q_i of any second order tensor $\mathbf{q}$ in such a way that $q_1 \geq q_2 \geq q_3$. The principal axes of the tensor are then named after the index of the principal value they refer to. There is no compelling reason to do this though. Clearly, different conventions will imply different angles to be associated to a given value of index i, which would make the proposed evolution rule ambiguous, to say the least.

In order to avoid any ambiguity in handling angles α_i in practical calculations, we shall assume that the principal axes of $\dot{\mathbf{e}}$ and $\boldsymbol{\varepsilon}^\circ$ are numbered according to the following convention: the principal axis of $\dot{\mathbf{e}}$ and the principal axis of $\boldsymbol{\varepsilon}^\circ$ which are at the smallest angle with each other will be called respectively axis 1 of $\dot{\mathbf{e}}$ and axis 1 of $\boldsymbol{\varepsilon}^\circ$. Accordingly, the angle between them will be denoted by α_1. Likewise, index $i = 2$ will be reserved to the principal axes of $\dot{\mathbf{e}}$ and $\boldsymbol{\varepsilon}^\circ$ that form the second smallest angle, which will then be α_2. Finally, index $i = 3$ will be given to the remaining two axes of the above tensors, whose angle will therefore be α_3. Clearly, with this convention, we shall have $\alpha_1 \leq \alpha_2 \leq \alpha_3$. Of course, the index to be associated to the principal values of $\dot{\mathbf{e}}$ and $\boldsymbol{\varepsilon}^\circ$ will have to be the same as the one we gave to the corresponding principal axes. This may produce a departure from the standard ordering convention of the principal values of these tensors, which of course has to be carefully reckoned within the calculations.

The above convention has a precise physical meaning as far as the evolution rule of $\boldsymbol{\varepsilon}^\circ$ is concerned. This is better seen by observing that, no matter the value of $\boldsymbol{\varepsilon}^\circ$, the principal triad of $\dot{\mathbf{e}}$ can be directed as we like because it depends on the strain we apply to the material. However, in view of eq. (7.2.1), the time rate $\dot{\varepsilon}_i^\circ$ of principal strain ε_i° depends on the principal value of $\dot{\mathbf{e}}$ whose principal direction is the nearest one to that of ε_i°. In other words, as far as the evolution of $\boldsymbol{\varepsilon}^\circ$ is concerned, it is the direction of the principal values of $\dot{\mathbf{e}}$ and not their magnitude that matters.

Once the label of the principal directions of tensors $\dot{\mathbf{e}}$ and $\boldsymbol{\varepsilon}^\circ$ is assigned, the arrows in each principal triad of axes can be assigned according to the right-hand or the left-hand sequence as needed to make them consistent with the triad of coordinate axes.

For practical applications, it may be more convenient to express the proposed evolution law in terms of a tensor $\mathbf{D}$ that rotates $\gamma\dot{\mathbf{e}}$ into $\dot{\boldsymbol{\varepsilon}}^\circ$.

The components of **D** can be determined as described in the last part of Appendix B, leading to eq. (B.40). Though a bit lengthy, that procedure does not present serious difficulties. The particular case in which a principal axis of $\boldsymbol{\varepsilon}^\circ$, say the one corresponding to $i = 3$, coincides with the homologous axis of $\dot{\mathbf{e}}$ occurs frequently in practice when dealing with plane states of stress. In this case, the tensor **D** can be determined directly from eq. (B.42), once angle θ_3 between the first principal directions of $\boldsymbol{\varepsilon}^\circ$ and $\dot{\mathbf{e}}$ is known.

In terms of tensor **D**, the evolution law (7.2.1) can be expressed as:

$$\dot{\boldsymbol{\varepsilon}}^\circ = \begin{cases} \gamma(\boldsymbol{\xi})\mathbf{D}^{\mathrm{T}}\dot{\mathbf{e}}\mathbf{D} & \text{if } f = 0 \text{ and } \dfrac{\partial f}{\partial \sigma_{ij}}\dot{\sigma}_{ij} > 0 \\ 0 & \text{if } f < 0 \text{ or } f = 0 \text{ and } \dfrac{\partial f}{\partial \sigma_{ij}}\dot{\sigma}_{ij} \leq 0. \end{cases} \tag{7.2.3}$$

Expressed in words, this law states that for purely elastic processes the rate of change of $\boldsymbol{\varepsilon}^\circ$ should vanish, while for elastic-plastic processes it is determined by the rate of change of the deviatoric part $\dot{\mathbf{e}}$ of total strain. When not vanishing, $\dot{\boldsymbol{\varepsilon}}^\circ$ will therefore be obtained from $\dot{\mathbf{e}}$ by first reducing it by factor γ and then rotating the resulting tensor by **D**. The latter operation will of course produce a bias of the principal directions of $\dot{\boldsymbol{\varepsilon}}^\circ$ toward the principal directions of $\boldsymbol{\varepsilon}^\circ$.

7.3 DRAG AND REDUCTION FACTORS IN PRACTICE

The evolution law of the previous section does not introduce any restriction on the possible form of functions $r = r(\boldsymbol{\xi})$ and $\gamma = \gamma(\boldsymbol{\xi})$. These functions should in principle be determined by experiment, since they represent constitutive quantities. Accordingly, their expression as well as the kind and number of variables $\boldsymbol{\xi}$ which they depend upon is expected to be different for different materials. Leaving $r(\boldsymbol{\xi})$ and $\gamma(\boldsymbol{\xi})$ unspecified, however, would lead to a far more general formulation than the one usually needed in practice. For this reason we shall presently confine our attention to the special case in which these functions reduce to:

$$r = \text{constant} \tag{7.3.1}$$

and

$$\gamma = \gamma(\omega \cdot w^p). \tag{7.3.2}$$

Here, w^p is the plastic work done during the last plastic loading process, while ω is a dimensionless *work-hardening factor*, usually greater than one, which keeps account of the effect of the *accumulated plastic work*. The latter will be denoted by $\tilde{w}$ and defined as the plastic work that is done on the material during its deformation history starting from the virgin state up to the beginning of the last plastic loading process. From that time on, $\tilde{w}$ will remain fixed to its last value, while any further plastic work will increase w^p.

Factor ω, appearing in the above equation, is supposed to be a function of $\tilde{w}$, i.e.

$$\omega = \omega(\tilde{w}). \tag{7.3.3}$$

The actual form of this function will of course be different for different materials. We shall assume that $\omega = 1$ if the material has no previous plastic deformation history, that is, if we are considering the first plastic loading of a deformation process that started from the virgin state. In this case, it is clear that $\tilde{w} = 0$, and the above assumption implies that $\omega(0) = 1$. For monotonic loading processes starting from the virgin state, therefore, function (7.3.2) will reduce to $\gamma = \gamma(w^p)$. The physical motivation of introducing ω into (7.3.2) will be discussed in Sect. 7.6.

Though particular, the above constitutive assumptions are nonetheless capable of describing many experimental findings concerning the subsequent yield surfaces of ductile materials of practical interest. In assessing them, one should not forget that they are intended to apply to the whole range of deformation processes that a material can ever suffer, no matter the dimensions of the applied stress nor the extent of plastic deformation. As such, they are an attempt at modelling by means of a few simple functions a phenomenon that is very complicated indeed, as it strongly depends on the deformation history of the material.

7.4 REMARKS ON THE EXPERIMENTAL DETERMINATION OF γ

An uniaxial test in the elastic-plastic range is the easiest – though not necessarily the most accurate – means to determine the values of reduction factor γ during an elastic-plastic process. A convenient feature of this test is that the strain increments $\dot{\boldsymbol{\varepsilon}}$ (and thus $\dot{\mathbf{e}}$) are always coaxial with the uniaxial state of strain of the specimen. A further simplification is obtained when the process starts from a specimen in virgin conditions. This makes $\boldsymbol{\varepsilon}^\circ = \mathbf{0}$ initially, which means that $\boldsymbol{\varepsilon}^\circ$ will be trivially coaxial with the deviatoric strain increment $\dot{\mathbf{e}}$ of the specimen as the first plastic deformation takes place.

Under these conditions, all angles α_i will vanish at the onset of the first plastic deformation. In view of eq. (7.2.2), this means that factor r becomes irrelevant for the computation of the first increment $\dot{\boldsymbol{\varepsilon}}^\circ$ of $\boldsymbol{\varepsilon}^\circ$. It also means that in order to calculate this increment, we must set $\mathbf{D} = \mathbf{1}$ in eq. (7.2.3) and that, moreover, the new value of $\boldsymbol{\varepsilon}^\circ$ too, i.e. $\dot{\boldsymbol{\varepsilon}}^\circ\,\mathrm{dt}$, will be coaxial with $\boldsymbol{\varepsilon}$.

As a result, the next deviatoric strain increment produced by the test will again be coaxial with the new value of tensor $\boldsymbol{\varepsilon}^\circ$: a situation that will repeat itself at each further strain increment. This will make $\mathbf{D} = \mathbf{1}$ throughout the test. When not vanishing, therefore, tensor $\dot{\boldsymbol{\varepsilon}}^\circ$ will be given by:

$$\dot{\boldsymbol{\varepsilon}}^\circ = \gamma\,\dot{\mathbf{e}}, \tag{7.4.1}$$

as immediately results from eq. (7.2.3). In this expression, γ denotes here the appropriate value assumed by $\gamma(\boldsymbol{\xi})$ at the time when the considered deformation increment takes place. For sufficiently small time increments Δt, increment $\Delta\boldsymbol{\varepsilon}^\circ$ of $\boldsymbol{\varepsilon}^\circ$ can be calculated from eq. (7.4.1) as given by:

$$\Delta\boldsymbol{\varepsilon}^\circ = \gamma\,\Delta\mathbf{e}. \tag{7.4.2}$$

This equation determines the value of γ once $\Delta\mathbf{e}$ and $\Delta\boldsymbol{\varepsilon}^\circ$ are known. $\Delta\mathbf{e}$ is immediately obtained from total strain increment $\Delta\boldsymbol{\varepsilon}$, since it is the deviatoric part of it. The procedure to obtain $\Delta\boldsymbol{\varepsilon}^\circ$ is more elaborate and is discussed below.

A further property of the considered uniaxial test is that in this case, tensor $\boldsymbol{\varepsilon}^\circ$ is completely determined by its principal values ε_i°. This is so because the principal directions of $\boldsymbol{\varepsilon}^\circ$ coincide with those of the total strain tensor applied to the specimen, which are known and fixed throughout the test. To be more definite, we shall take principal direction 1 of $\boldsymbol{\varepsilon}^\circ$ as the one that parallels the specimen axis. The other two directions will, accordingly, be normal to that axis. Their principal values, namely ε_2° and ε_3°, must be the same, as immediately results from simple symmetry arguments. Moreover, by remembering that $\boldsymbol{\varepsilon}^\circ$ is assumed to be deviatoric (cf. Sect. 2.3), we can conclude that:

$$\varepsilon_2^\circ = \varepsilon_3^\circ = -\frac{1}{2}\varepsilon_1^\circ, \tag{7.4.3}$$

ε_1° being the principal value of $\boldsymbol{\varepsilon}^\circ$ relevant to principal direction 1. In the considered process, therefore, $\boldsymbol{\varepsilon}^\circ$ is fully determined by its principal component ε_1°.

On the other hand, the value of ε_1° can be determined by exploiting the yield condition. Take a coordinate system (x_1, x_2, x_3) whose x_1-axis coincides with the specimen axis, and denote by σ the axial stress of the specimen. The elastic strain tensor will then be given by:

$$\boldsymbol{\varepsilon}^e = \begin{bmatrix} \varepsilon_1^e & 0 & 0 \\ 0 & -\nu\varepsilon_1^e & 0 \\ 0 & 0 & -\nu\varepsilon_1^e \end{bmatrix} = \begin{bmatrix} \frac{\sigma}{E} & 0 & 0 \\ 0 & -\nu\frac{\sigma}{E} & 0 \\ 0 & 0 & -\nu\frac{\sigma}{E} \end{bmatrix}. \qquad (7.4.4)$$

The deviatoric component of this tensor is:

$$\mathbf{e}^e = \begin{bmatrix} \frac{2}{3}(1+\nu)\frac{\sigma}{E} & 0 & 0 \\ 0 & -\frac{1}{3}(1+\nu)\frac{\sigma}{E} & 0 \\ 0 & 0 & -\frac{1}{3}(1+\nu)\frac{\sigma}{E} \end{bmatrix}. \qquad (7.4.5)$$

We can now introduce this tensor into eq. (6.1.1) (remember that the notation e_i that we adopted in writing eq. (6.1.1) was in fact shorthand for e_i^e). In doing so we should set $\mathbf{Q} = \mathbf{R}^\circ = \mathbf{1}$, since in the present case both $\boldsymbol{\varepsilon}^\circ$ and $\boldsymbol{\varepsilon}^e$ are coaxial with the reference axes. By making use of eqs (5.2.5) and (7.4.3) and by recalling that $G/E = 1/[2(1+\nu)]$, from eq. (6.1.1) we thus get:

$$\sigma_y^2 - 6\,G\,\sigma_y\,\varepsilon_1^\circ = 6\,G\,(\kappa^2 - b). \qquad (7.4.6)$$

Here σ_y is the axial yield stress in tension, corresponding to the current state of plastic deformation of the specimen.

Due to the presence of the unknown quantity b, eq. (7.4.6) is not sufficient to determine ε_1°. A further independent equation is therefore needed. To obtain it, we can unload the specimen and then reload it elastically along a different stress path until a different yield point on the same yield surface is reached. Such an unloading and subsequent reloading occur entirely in the elastic range, which leaves the value of $\boldsymbol{\varepsilon}^\circ$, and thus of ε_1°, unchanged. As for the new point on the yield surface, any point will do, provided that it is different from the yield point in simple tension already considered in the first part of the process. If we want to stick to the uniaxial test, the only choice we have is to reload the specimen in compression until the yield limit in simple compression, say $\overline{\sigma}_y$, is reached (the value of $\overline{\sigma}_y$ is of opposite sign to that of σ_y , if the stress-free state of the material falls within the elastic region). A similar calculation as the one leading to eq. (7.4.6) will in this case produce:

$$\overline{\sigma}_y^2 - 6\,G\overline{\sigma}_y\varepsilon_1^\circ = 6\,G(\kappa^2 - b). \qquad (7.4.7)$$

This equation together with eq. (7.4.6) determines the values of b and ε_1° relevant to the considered value of w^p. The corresponding value of γ will then follow immediately from eq. (7.4.2).

Admittedly, the elastic limit in compression may in practice be difficult to determine accurately, especially so if we use the same slender specimen that is used for the tensile test. A possible alternative could be to elastically reload the specimen in pure shear rather than in simple compression. This could easily be done by exploiting the experimental apparatus that is used to test thin-walled tubular specimens in tension and torsion. The yield stress in pure shear will then be attained by loading the specimen in pure torsion. The limit curves relevant to tension/torsion tests are given by eq. (6.4.12). In order to apply this equation to the present state of stress, we must set $\sigma = 0$ and $\tau = \bar{\tau}_y$ in it. Here, $\bar{\tau}_y$ indicates the yield stress in pure shear, as obtained by elastically reloading the specimen in pure torsion after unloading it from the previous loading in simple tension. If the axis of the tubular specimen coincides with the x_1-axis, condition $\mathbf{R}^\circ = \mathbf{1}$ still applies, which makes $\beta = 0$ in eq. (6.4.12). In this case, however, $\theta_3 = \pi/4$ rad, as follows from eq. (6.4.2) for $\sigma = 0$. From eqs (6.4.12), (5.2.5), (7.4.3) and (A.14), we would then finally get:

$$\bar{\tau}_y^2 + \frac{2}{\sqrt{2}} G\,\bar{\tau}_y\,\varepsilon_1^\circ = 2\,G(\kappa^2 - b), \tag{7.4.8}$$

which again, together with eq. (7.4.6) determines both ε_1° and, from eq. (7.4.2), the value of γ relevant to the considered value of w^p.

7.5 THE EXPERIMENTAL DETERMINATION OF r

The determination of drag factor r requires a different approach. The uniaxial test does not produce any rotation between $\boldsymbol{\varepsilon}^\circ$ and $\dot{\mathbf{e}}$ and, therefore, no angular drag can be produced. The tension/torsion test of thin-walled tubes can be used instead. In this case the principal directions of the elastic-plastic strain increment $\dot{\mathbf{e}}$ can be varied during the test, thus producing values of $\dot{\boldsymbol{\varepsilon}}^\circ$ that are not coaxial with $\boldsymbol{\varepsilon}^\circ$. The simplest way to proceed could be the following.

First of all, a plastic deformation process in simple tension is imposed on the specimen and the relevant value of $\boldsymbol{\varepsilon}^\circ$ is determined as described in the previous section. We then apply a torque to the specimen and make it suffer a further elastic-plastic deformation increment, say $\dot{\boldsymbol{\varepsilon}}$, whose value is known and controlled by the experimenter through torque and axial force applied at the ends of the specimen. Let $\dot{\mathbf{e}}$ be the deviatoric part of this increment. The principal directions of $\dot{\mathbf{e}}$ are known, since they coincide with the principal directions of the strain increment we impose on the specimen. Increment $\dot{\boldsymbol{\varepsilon}}^\circ$ can therefore be obtained by applying the evolution law of Sect. 7.2. To do so, however, we must assign a value to r, which is unknown at the outset, since it is the quantity we want to determine.

To overcome this difficulty, we observe that the above procedure will produce different values of $\dot{\boldsymbol{\varepsilon}}^\circ$ for different values of r. Hence, depending on the value we assume for r, different new values of $\boldsymbol{\varepsilon}^\circ$ will be calculated corresponding to the same plastic strain increment imposed on the specimen. For each value of r, eq. (6.4.12) will give a different yield curve in the σ/τ plane. The value of b in that equation is determined by requiring that the yield curve corresponding to the considered value of r should pass through the stress point at which the last plastic deformation occurred. The right value of r can then be obtained by choosing from among all such yield curves the one that best fits the other experimentally determined points of the actual σ/τ yield curve of the material.

Of course, such a procedure requires that an appropriate number of experimental points of the actual yield curve should be known beforehand. It also requires that the value of γ should be given in advance. This is not a problem, as γ is independently determined by the procedure in the previous section.

7.6 QUALITATIVE APPRAISAL OF γ FROM UNIAXIAL TESTS

Constitutive assumptions (7.3.2) and (7.3.3) mean that the functional dependence of γ on ω and w^p is the same, no matter the process. In principle, therefore, complete information on $\gamma(\omega \cdot w^p)$ could be obtained by confining our attention to a conveniently simple family of processes. The family of uniaxial stress processes could be chosen to this end, as these processes are easily performed on standard specimens on standard test machines. One should not forget, however, that assumptions (7.3.2) and (7.3.3) are meant to be an approximation – possibly a good one, but still an approximation – of a more complicated functional dependence of γ on a greater number of variables that better specify its dependence on plastic deformation history. This suggests that the experimental determination of function $\gamma(\omega \cdot w^p)$ should be better sought by taking appropriate averages of the values that, for given ω and w^p, the reduction factor assumes in a wide range of processes of different kinds. A rough idea on the shape of this function can, however, be obtained quite easily from the ubiquitous σ/ε curve relevant to standard uniaxial tests in the plastic range, as discussed below.

The first observation to be made is that γ controls the rate of change of tensor $\boldsymbol{\varepsilon}^{\circ}$, which is the main responsible for the changes of the yield surface following plastic deformation. This gives us an important clue about the values that γ should assume during a plastic deformation process. Take the case of elastic-perfectly plastic materials, for instance. The elastic limit of these materials is not affected by plastic deformation, which implies flat uniaxial σ/ε curves in the plastic range. A flat uniaxial σ/ε curve in the plastic range indicates that there is no change in the elastic limit and that, therefore, γ must vanish no matter the value of w^p. Contrast this with the uniaxial σ/ε curves of work-hardening materials. These curves exhibit a non-vanishing slope in the plastic range, meaning that the elastic limit depends on plastic deformation. In work-hardening materials, therefore, γ should assume appropriate non-vanishing values, generally different for different values of w^p.

In order to better illustrate the relationship between the slope of uniaxial σ/ε curves and the shape of function $\gamma(\omega \cdot w^p)$, let us refer to the particular case of a material in which b is equal to zero. From eqs (7.4.6) and (7.4.1) it follows that, in this case, σ_y depends on ε_1° or, equivalently, on γ only. In the case of uniaxial tests, the axis of the specimen is a principal axis for stress and strain. Let us for the moment confine our attention to processes that make the deformation of the specimen increase monotonously from the initial stress-free virgin state. As already observed, in this case we have $\tilde{w} = 0$ and hence $\omega = 1$, which reduces eq. (7.3.2) to $\gamma = \gamma(w^p)$. If the slope of the σ/τ curve is not too steep, the value of w^p can be approximated to the area under this curve, starting from the point A where the first plastic yielding takes place (see inset of Fig. 7.6.1).

Consider then a generic point P in the elastic-plastic part of the σ/ε curve and suppose that the slope at this point of the curve is, say, positive (see

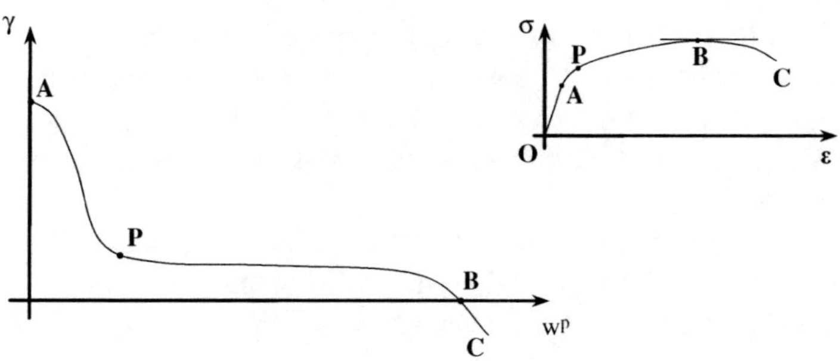

Fig. 7.6.1 A standard uniaxial σ/ε curve (inset) permits a qualitative estimate of $\gamma(w^p)$ for monotonic uniaxial plastic loading processes starting from the virgin state.

inset of Fig. 7.6.1). Let w^p be the value of plastic work accumulated during the inelastic part of the process from point A to point P. Since the slope of the σ/ε curve is positive at P, a further increase in the deformation of the specimen will produce an increase in the elastic limit of the material. In view of eq. (7.4.6), this requires that ε_1° should increase too, which from eq. (7.2.1) implies that at the considered point, (i.e. for the considered value of w^p) factor γ should be positive. The steeper the σ/ε curve at P, the greater the value of γ relevant to w^p.

Likewise, γ must be negative at the points of negative slope; the more negative the slope, the lesser the value of γ. Finally, consistent to what we observed above when referring to perfectly plastic materials, γ must vanish at the points of vanishing slope of the σ/ε curve (cf. point B in inset of Fig. 7.6.1). From these considerations, a qualitative estimate of function $\gamma(w^p)$ can be easily obtained once the uniaxial σ/ε curve of the material is given. This is illustrated in Fig. 7.6.1.

Crucial to the above arguments is the assumption that $b \equiv 0$. In general, b will be different from zero, though. This can produce considerable distortions in the above qualitative curve of $\gamma(w^p)$, since the subsequent elastic limits of the material will not then be controlled by γ alone. In this case, a precise evaluation of function $\gamma(w^p)$ can of course be obtained from eq. (7.2.1), once ε_1° and b are calculated as described in Sect. 7.2.

Reference to uniaxial test also helps in justifying the introduction of factor ω into eq. (7.3.2). To see this, let us consider the case in which the specimen is unloaded from a point in the elastic-plastic part of the σ/ε curve and then reloaded beyond the new elastic limit. A number of such reloading processes are represented in Fig. 7.6.2. They refer to different plastic deformation histories, which means different values of $\tilde{w}$ for each history. More specifically, the reloading paths that are further to the right of

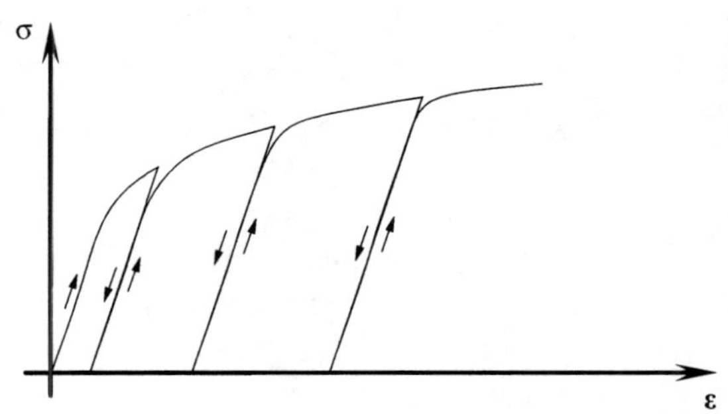

Fig. 7.6.2 Reloading deformation processes that start from larger plastic deformation have sharper elastic-plastic transitions.

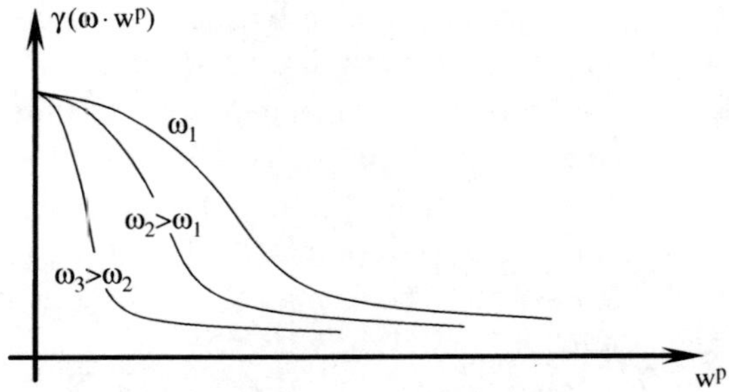

Fig. 7.6.3 Influence of ω on γ [cf. eq. (7.3.2)]: as ω is increased, γ changes more quickly with w^p.

the diagram correspond to larger previous plastic deformations and hence to larger values of $\tilde{w}$. The same diagram shows that as past plastic deformation is increased, the elastic-plastic transition upon reloading becomes sharper. This is a well-known experimental fact. It suggests that in these transitions, the value of γ should change more rapidly in specimens that suffered larger values of previous plastic deformation. A way to model this is to multiply the independent variable w^p entering function $\gamma(w^p)$ by a factor ω whose actual value depends on $\tilde{w}$. This is exactly what the combination of eqs (7.3.2) and (7.3.3) does.

Figure 7.6.3 illustrates the effect that an increase of work-hardening factor ω has on γ.

7.7 REVERSE LOADING AND THE INFUENCE OF THE LOADING DIRECTION

Plastic loading following load reversal from a previous plastic loading in the opposite direction will be referred to as reverse loading. A reverse loading process exhibits some particular features that any realistic work-hardening rule should model. Here, we shall confine our attention to reverse loading under uniaxial stress, because experimental knowledge on loading reversal under more general states of stress is scanty.

A well-established experimental feature of these processes is that no matter how strongly we plastically strain a specimen uniaxially, say in tension, the elastic-to-plastic transition at the first load reversal (which would then be in compression) is quite gradual. In many materials the shape of the σ/ε curve corresponding to the first load reversal is similar – though in the opposite direction – to that of the first loading from the virgin state, irrespective of how large the previous plastic strain was. For further cycles of removal and re-application of the reverse load, however, this transition becomes sharper and sharper, thus mimicking in compression the behaviour that the material exhibits under repeated plastic loading in tension. The situation is depicted in Fig. 7.7.1.

The main point to be noted here is that the elastic-plastic transitions in the above two opposite directions are largely independent of each other. Neither the number of transitions suffered by the material in one given direction, nor the extent of plastic strain they involve appear to affect the behaviour of the material in the opposite direction. This suggests that, in applying eqs (7.3.2) and (7.3.3), we should keep two separate accounts of the accumulated plastic work $\tilde{w}$ in tension and in compression. Two generally different values of $\tilde{w}$ should thus be used when calculating the evolution of the yield surface of the material following a plastic deformation process, in tension and in compression.

Admittedly, this feature reveals a weakness of assumptions (7.3.2) and (7.3.3), in that it requires that the reduction factor γ should depend on the loading direction. It must be acknowledged, however, that the available experimental data on the effect that plastic strain in one direction has on the evolution law of the yield surface in a different direction is near to naught. For this reason, the reader should be well advised that the proposed evolution rule may work fairly well for processes where the strain direction does not suffer major changes or, possibly, when it varies smoothly enough.

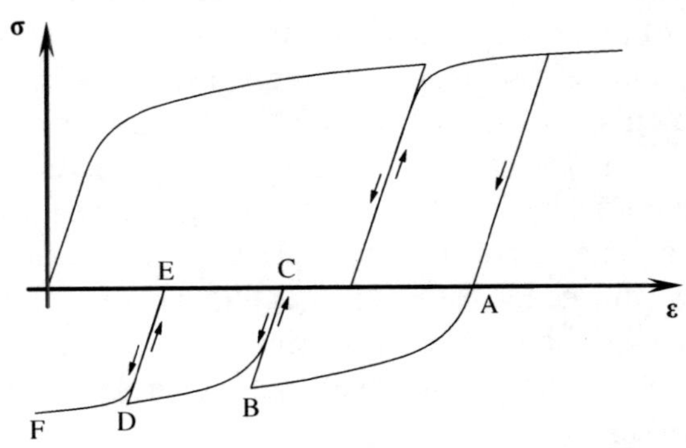

Fig. 7.7.1 No matter how strong the plastic deformation in tension, the first elastic-plastic transition in compression (path AB in the figure) is roughly as gradual as the first tensile transition of the virgin material. For repeated reloading in compression, the transition becomes sharper and sharper as the plastic strain in compression increases (cf. paths CD and EF in the figure).

While waiting for more experimental work on this topic, eqs (7.3.2) and (7.3.3) can be applied to uniaxial processes involving loading reversals provided that two separate accounts for past plastic work $\tilde{w}$, in tension and in compression, are taken. For a more general discussion on the way in which the influence of past strain direction can be introduced into the evolution equations of the yield surface, the reader is referred to Appendix C of this book.

7.8 EVOLUTION RULE FOR d_o

Before attempting to formulate any particular constitutive assumption for d_o, it is convenient to recall here its physical meaning. In Sect. 5.1 we defined $\Delta\overline{\psi}$ as the elastic energy that is entrapped in the matrix due to elastic distortions caused by microscopic defects. The part of $\Delta\overline{\psi}$ that is accommodated in the matrix without producing any macroscopic strain was then denoted as $\overline{c}$. Finally in Sect. 5.2, the deviatoric part of $\overline{c}$ was denoted by d_o. The latter is, therefore, the deviatoric part of the elastic energy entrapped in the matrix through a system of microscopic elastic distortions that does not result in any macroscopic strain.

None of the above-mentioned quantities suffers any change in value if the material is kept in its actual elastic range. For this reason, in the previous sections they were often referred to as constants. Their value, however, depends on the defect content of the material and hence, ultimately, on the plastic deformation process it suffered. As observed in Sect. 7.1, d_o enters the expression of the subsequent yield surfaces of the material through b. The law of its evolution with plastic deformation must be known in advance if we want to predict how the yield surface will be modified by the elastic deformation process. The problem is a constitutive one. As such, its answer must be sought on the basis of experimental data. To begin with, in trying to find the appropriate law relating d_o to plastic deformation, it may help to formulate a few educated guesses. In what follows we shall propose some. It should be stressed, however, that the right evolution law for d_o, as well as the variables upon which d_o may depend, will in general be different for different materials.

7.8.1 Proposal A: No evolution at all

This is the simplest evolution rule of all. It amounts to stating that:

$$d_o = \text{constant} \tag{7.8.1}$$

no matter the plastic deformation process of the material. If $d_o \neq 0$, this law is clearly inapplicable at the beginning of a plastic deformation processes starting from the virgin material. The reason is that a constant non-vanishing value of d_o requires an initial plastic deformation process to make d_o reach such a value; which clearly contradicts assumption (7.8.1). It may well be, though, that in such a process d_o builds up quickly to its final value (7.8.1). In this case eq. (7.8.1) will in practice apply to every plastic deformation process, if we only exclude a short initial process that starts from the virgin state. Such a situation may not be unreasonable, since it is likely that the amount of elastic energy that a material can accumulate without suffering macroscopic strain is quite limited and hence saturates quickly. As a matter of fact, eq. (7.8.1) turns out to be consistent with the experimental results of Phillips and Lu [9] that will be discussed in Ch. 9. It should be acknowledged, however, that this fact alone is not enough to draw any general conclusion about the range of applicability of this proposal.

7.8.2 Proposal B: Elastic region of constant size

The analytical form of the general expressions of the complete (six-dimensional) yield surface that we presented in Sect. 6.1 reveals that the size of the elastic region varies homotetically as b is changed. Quantities a_i and $\mathbf{R}^\circ$ affect the position and the orientation of this surface, but not its size. Accordingly, if we assume that the initial virgin elastic region moves rigidly as the plastic deformation proceeds, we must assume that $b \equiv 0$. In view of eq. (5.2.6), this means that:

$$d_o = d_o(\varepsilon^\circ) = -\mu(\varepsilon_1^{\circ 2} + \varepsilon_2^{\circ 2} + \varepsilon_3^{\circ 2}). \tag{7.8.2}$$

In the light of the above observations, such an assumption would represent a sort of kinematic work-hardening rule. It must be noted, however, that such an interpretation applies to the full six-dimensional yield surface and not to the elastic sub-domains that result when we keep some components of stress or elastic strain constant. For instance, eq. (7.8.2) by no means entails a kinematic work-hardening rule for the σ/τ yield curves as obtainable from tension/torsion tests of thin-walled tubes.

7.8.3 Proposal C: Constant ratio hypothesis

This is a generalization of the previous case. It is assumed that the part of deviatoric elastic energy that is entrapped in the matrix at no macroscopic strain is proportional to the part of entrapped energy that is due to permanent elastic strain $\boldsymbol{\varepsilon}^{\circ}$. That is:

$$d_o = d_o(\varepsilon^{\circ}) = \Theta(\varepsilon_1^{\circ 2} + \varepsilon_2^{\circ 2} + \varepsilon_3^{\circ 2}), \tag{7.8.3}$$

Θ being a suitable constant. A further generalization of this rule would make Θ depend on plastic work or on any other variable related to plastic deformation.

Of course, an unlimited number of other constitutive assumptions can be formulated. The ones we presented above are but the simplest ones. The available experimental data on subsequent yield surfaces do not seem to allow for more definite conclusions at the present.

7.9 PLASTIC FLOW EQUATIONS, CONVEXITY OF THE YIELD SURFACE AND NORMALITY RULE

When regarded as a function of $\{e_1, e_2, e_3, \mathbf{Q}\}$, the general yield function (6.1.1) represents a surface in space of variables e_i and $\mathbf{Q}$. As already observed, such a surface is a hyper-surface in a space of six dimensions, since tensor $\mathbf{Q}$ is defined by just three Euler's angles. A welcome feature of this surface is that it is convex. This can readily be proved by noting that *any* cross-section at constant $\mathbf{Q}$ of it is convex; in fact it is an ellipsoid. The convexity of the yield surface and normality of the plastic flow are needed for a material to be stable in Drucker's sense.

Unless otherwise stated, we shall assume that the material is stable. In this case, eq. (1.2.20) is a suitable plastic flow rule, in that it provides that the plastic flow should be normal to the surface (6.1.1) – no matter how strongly the material has been work-hardened.

In considering the above properties of convexity and normality, care must be exercised, however, in making reference to the whole yield surface in its full six- dimensional space and not to limit curves or limit surfaces of reduced dimensions resulting from particular cross-sections of it. If the yield function is anisotropic, the chances are that these limit curves or sub-surfaces are non-planar cross-sections of the complete yield surface (6.1.1). As a consequence, when represented in a flat space of reduced dimensions they may not be convex, as already remarked in Sect. 1.3, nor will the normality rule apply to them.

The case of the σ/τ yield curves defining the elastic domains from standard tension/torsion experiments provides a typical instance of this. As discussed in Sect. 1.3, plastic limit anisotropy due to cold work may make these curves non-convex in spite of the fact that the complete yield surface to which they belong is convex. In this case, the lack of convexity does not indicate that the material is unstable. It is simply the geometric consequence of representing in a plane a limit curve that, in fact, is a non-planar curve on an otherwise convex surface in a space of higher dimensions. That strongly cold worked materials may exhibit non-convex σ/τ yield curves is a fact that is not lacking some – if seldom acknowledged – experimental evidence, as we shall discuss later in this book.

8

Flat Bars and Thin-Walled Tubes after Uniaxial Plastic Straining

Oddly enough, after uniaxial plastic prestraining, the σ/τ yield curve as obtained from a specimen of material in the form of a flat bar or a thin-walled tube may differ from that of an element of the same material subjected to the same prestraining history. This is caused by a combination of two factors. One is related to the rise of anisotropy in the yield surface due to plastic deformation. The second is due to the fact that, in the plastic range, the motion (not the strain!) of the points of these specimens is only partially controlled by the displacement of the specimen ends. As shown in the next section, two alternative motions can – and generally will – take place at different points of such specimens during the same uniaxial strain process. Since these motions differ from each other by a rigid-body rotation, both of them generate the same strain history at every point of the specimen, thus leading it to the same state of uniform stress and strain. Their different rotation, however, will in general result in different yield surfaces at different points of the specimen if the yield surface of the material is not isotropic, as usually happens after plastic straining.

The consequence is that, after uniaxial prestraining, the overall σ/τ yield curve of a thin-walled tube or a flat bar may differ from the local yield surface predicted by eq. (6.4.12). In the present chapter, we shall study the details of this phenomenon and discuss its relevance in interpreting the experimental findings obtained from standard tension/torsion experiments on thin-walled tubes.

For simplicity's sake, we shall consistently refer to flat bars rather than thin-walled tubes. The equivalence of the two kinds of specimens as far as N/T tests are concerned was pointed out in Sect. 6.4. The reference to flat bars will give us the advantage of an easier description of stress and strain in ordinary Cartesian coordinates. In practice, however, tests under combined axial and shear stress are far more easily made by applying axial force and torque to the ends of a thin-walled tube rather than by applying axial and shearing forces to the boundaries of a flat bar.

8.1 DIFFERENT MOTIONS DURING THE SAME TEST

Let a flat bar of an elastic-plastic material be subjected to a simple uniaxial tensile test, as shown in Fig. 8.1.1a. Both stress and strain are uniform throughout the bar during the process, whether the process is purely elastic or not. Their values are controlled by axial force and axial displacement at the ends of the bar. In the Cartesian coordinate system shown in Fig. 8.1.1, the only non-vanishing component of stress is $\sigma_{11} \equiv \sigma = N/A$, the quantity A being the area of the bar cross-section normal to axis x_1. The deviatoric component of stress is therefore:

$$\mathbf{s} = \begin{bmatrix} \frac{2}{3}\sigma & 0 & 0 \\ 0 & -\frac{1}{3}\sigma & 0 \\ 0 & 0 & -\frac{1}{3}\sigma \end{bmatrix}. \tag{8.1.1}$$

Suppose now that the elastic limit is exceeded and that the tensile force is increased to produce a certain amount of plastic strain in simple tension.

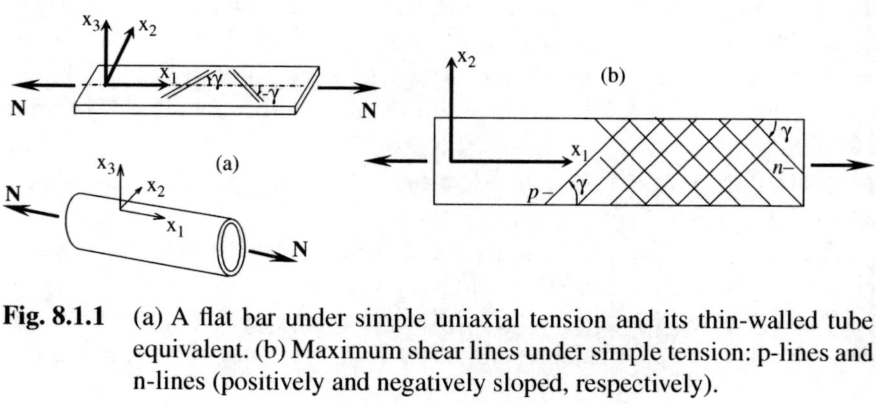

Fig. 8.1.1 (a) A flat bar under simple uniaxial tension and its thin-walled tube equivalent. (b) Maximum shear lines under simple tension: p-lines and n-lines (positively and negatively sloped, respectively).

In order to enquire about the plastic displacement field that results from such a prestraining, let us decompose $\mathbf{s}$ into two states of pure shear stress according to the relation:

$$\mathbf{s} = \mathbf{s}' + \mathbf{s}'', \tag{8.1.2}$$

where

$$\mathbf{s}' = \begin{bmatrix} \frac{2}{3}\sigma & 0 & 0 \\ 0 & -\frac{2}{3}\sigma & 0 \\ 0 & 0 & 0 \end{bmatrix} \quad \text{and} \quad \mathbf{s}'' = \begin{bmatrix} 0 & 0 & 0 \\ 0 & \frac{1}{3}\sigma & 0 \\ 0 & 0 & -\frac{1}{3}\sigma \end{bmatrix}. \tag{8.1.3}$$

In view of eq. (1.2.19), this implies that the plastic strain $\boldsymbol{\varepsilon}^{\mathrm{p}}$ that is produced by the considered process can be decomposed into two states of simple shear as follows:

$$\boldsymbol{\varepsilon}^{\mathrm{p}} = \boldsymbol{\varepsilon}'^{\mathrm{p}} + \boldsymbol{\varepsilon}''^{\mathrm{p}}, \tag{8.1.4}$$

where

$$\boldsymbol{\varepsilon}'^{\mathrm{p}} = \begin{bmatrix} \varepsilon_1^{\mathrm{p}} & 0 & 0 \\ 0 & -\varepsilon_1^{\mathrm{p}} & 0 \\ 0 & 0 & 0 \end{bmatrix} \quad \text{and} \quad \boldsymbol{\varepsilon}''^{\mathrm{p}} = \begin{bmatrix} 0 & 0 & 0 \\ 0 & \frac{1}{2}\varepsilon_1^{\mathrm{p}} & 0 \\ 0 & 0 & -\frac{1}{2}\varepsilon_1^{\mathrm{p}} \end{bmatrix}. \tag{8.1.5}$$

The quantity $\varepsilon_1^{\mathrm{p}}$ introduced above denotes plastic strain along axis x_1.

8.1.1 Plastic motion leading to $\boldsymbol{\varepsilon}'^{\mathrm{p}}$

Let us first of all consider the motion of the points of the specimen leading to plastic prestrain $\boldsymbol{\varepsilon}'^{\mathrm{p}}$. Actually, there are two different motions or *modes,* which can produce the same prestraining $\boldsymbol{\varepsilon}'^{\mathrm{p}}$. One mode is a simple shearing motion along the lines of maximum shear at an angle of $\pi/4$ rad with the x_1-axis. We shall refer to it as the p-mode. The other mode will be referred to as the n-mode. It is similar to the p-mode but takes place along the maximum shear lines that are at an angle of $-\pi/4$ rad with the x_1-axis. Both kinds of maximum shear lines are represented in Fig. 8.1.1b.

More generally, if both axial and shearing forces are applied to the specimen, the lines of maximum shear will form different angles with the specimen axis. The actual values of these angles depend on the values of the applied forces. We shall again distinguish them into the p-lines and n-lines, according to whether they form a positive or a negative angle with the bar axis. The prefixes p and n will help to remind us of the sign of their slope.

The above two modes of plastic prestraining under uniaxial stress differ from each other by a rigid-body rotation. The easiest way to see this is to observe that a specimen rotation of π rad about the x_1-axis brings the n-lines to superimpose to the p-lines. Thus, by rigidly rotating the bar by π rad about the x_1-axis, the displacement field due to a shearing motion along one family of lines can be brought to coincide with the analogous motion along the other family of lines.

These two ways of achieving the same plastic prestraining $\boldsymbol{\varepsilon}'^{p}$ may not be equivalent as far as their effect on subsequent plastic yielding is concerned, though. Differing from each other by a rigid-body rotation, the two modes will generally bring the material into different elastic regions if its yield surface is not isotropic. This will make the σ/τ yield limit at any given point of the specimen depend on which prestraining mode is activated at that point during the prestraining process. Yet the plastic strain history is exactly the same at every point of the specimen, no matter which prestraining mode is activated at which point.

The local σ/τ yield curves following the above two prestraining modes can be calculated from eqs (6.4.12) and (6.4.2)–(6.4.5). To do this, we must first of all set $\beta = 0$ in that equation. This is a consequence of the fact that, according to the work-hardening rule we introduced in the previous chapter, the considered uniaxial prestraining produces a permanent elastic strain $\boldsymbol{\varepsilon}^{\circ}$ that is coaxial with the assumed reference axes. (We are assuming here that plastic prestraining takes place in an initially virgin material.) It should then be remembered that the considered reference axes are supposed to be embedded in the specimen and, therefore, rotate with it. It follows that a π rad rotation of the specimen about the x_1-axis inverts the positive sign for the shear forces applied to it when they are viewed in the rotated reference system. Thus, a positive shear becomes negative for the rotated specimen. This makes sense from the physical standpoint too, since we are rotating the specimen and not the forces that are applied to it; which effectively inverts the relative direction of the shear forces with respect to the specimen itself (see Fig. 8.1.2a).

In the light of the above remarks, the σ/τ yield curve of the rotated element can simply be obtained from the same eqs (6.4.12) and (6.4.2)–(6.4.5), once $(-\tau)$ is substituted for τ. Geometrically speaking, this means that the local σ/τ yield curves of the element before and after a π rad rotation about axis x_1 mirror each other through to the σ-axis.

The two σ/τ yield curves thus obtained are the local σ/τ yield curves of the material relevant to prestraining mode p and n, respectively. This is due to the fact that, as we already pointed out, a π rad rotation about axis

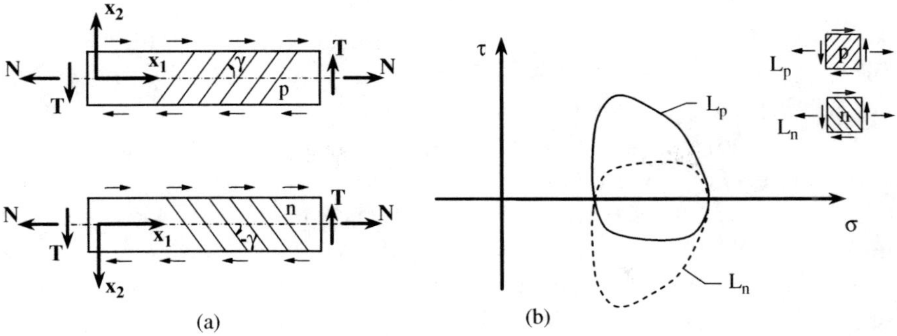

Fig. 8.1.2 (a) Specimen rotation of π rad about the x_1-axis changes the sign of the applied shear in the embedded coordinate system (x_1, x_2, x_3). (b) An instance of local σ/τ yield curves relevant to the p-mode (curve L_p) and the n-mode (curve L_n) uniaxial prestraining. These curves mirror each other with respect to the σ-axis.

x_1 makes one prestraining mode coincide with the other. Of course, which of the two curves corresponds to which mode depends on the convention we assume for the positive direction of τ. (It may be worth to remember, in passing, that in plane (σ, τ) the sign of τ is understood to be given according to Mohr convention. As is well known, this convention is different from the sign convention for stress tensor components.) We shall henceforth denote by L_p and L_n the σ/τ yield curves resulting from prestraining modes p and n, respectively (see Fig. 8.1.2b).

From simple continuity arguments, it follows that if a point on the (σ, τ)-plane belongs, say, to the L_p curve and represents the yield limit for plastic flow to occur along a p-line, then its neighbouring points on the same curve also refer to yield limits along a p-line. The angle of the latter, however, will be different depending on the value of σ and τ at the considered point.

Since the L_p curve is defined as the one that for $\tau = 0$ gives the yield limit along a p-line, it follows that every point of the same curve will represent a yield limit for plastic flow to occur along a p-line. The angle of the latter will vary from 0 to $\pi/2$ rad, depending on the value and sign of σ and τ. Likewise, curve L_n will define the yield limits of the material for plastic flow to occur along the n-lines.

8.1.2 Contribution to specimen motion due to $\boldsymbol{\varepsilon}''^{p}$

The motion component that produces strain $\boldsymbol{\varepsilon}''^{p}$ remains to be considered. This is a simple shear strain in the normal planes to the bar axis. As such, it does not produce any displacement along that axis. Similar to what applies to $\boldsymbol{\varepsilon}'^{\mathrm{p}}$, strain $\boldsymbol{\varepsilon}''^{\mathrm{p}}$ too can be produced by two different pure shearing displacement fields. In this case, however, both displacement fields are normal to axis x_1 and are respectively directed at angles of $\pi/4$ rad and $-\pi/4$ rad with the x_2-axis. What is important to observe is that for any given value of $\boldsymbol{\varepsilon}''^{\mathrm{p}}$, the amplitude of these displacements tends to zero as the thickness of the bar becomes smaller and smaller. It follows that for thin enough bars, the displacement field due to uniaxial plastic prestraining can to a good approximation be considered as caused by the displacement field that produces the plastic strain component $\boldsymbol{\varepsilon}'^{\mathrm{p}}$ only. (To be convinced about this, remember that $\boldsymbol{\varepsilon}''^{\mathrm{p}}$ represents an isochoric deformation and that it does not produce any motion in direction x_1. This implies that the two non-vanishing principal values of $\boldsymbol{\varepsilon}''^{\mathrm{p}}$ – representing the plastic elongations across the thickness of the bar and across its width – must be of the same order of magnitude but opposite in sign. If the thickness of the bar tends to zero, the requirement that its volume should remain constant means that the strain across its width should tend to zero too.)

In what follows we shall confine our attention to thin flat bars. Accordingly, the displacement field relevant to strain component $\boldsymbol{\varepsilon}''^{\mathrm{p}}$ will be neglected.

8.2 REASON FOR THE LACK OF SYMMETRY IN SHEAR OF THE LOCAL σ/τ YIELD CURVES FOLLOWING UNIAXIAL PRESTRAINING

Tension/torsion tests of thin-walled tubes show that the σ/τ yield curves from specimens that were plastically prestrained in simple tension are symmetric with respect to the σ-axis. In other words, for any given value of the applied axial force, the amplitude of the torque needed to bring the specimen to the yield limit is the same no matter the sign of the torque. Yet the σ/τ yield curves shown in Fig. 8.1.2 are patently unsymmetrical with respect to the σ-axis.

In order to solve this apparent inconsistency, we must first of all make it clear that the σ/τ yield curves of Fig. 8.1.2 are meant to represent the *local* response of the material, whilst the above quoted tension/torsion symmetry refers to the *overall* response of the specimen. Although the specimen is in a uniform state of strain and stress throughout the test, such a state is generally produced by different motions in different parts of the specimen. These are the p- and n-modes we introduced in the previous section. As already remarked, they can produce different elastic regions if the yield function of the material becomes anisotropic due to plastic straining. In this case, the tension/torsion yield limits obtained from the test are in fact the overall response of a specimen that possesses different elastic limits at different points, in spite of the fact that both strain and strain history are homogeneous all over it.

By contrast, the local σ/τ yield curves L_p or L_n can be thought of as ideally obtained by making a σ/τ test on an infinitesimal bar element. For such an element, the prestraining process is trivially homogeneous (we cannot distinguish different points within an infinitesimal element!). Its σ/τ yield curves will then be either the L_p or the L_n one, depending on whether the p-mode or the n-mode is activated in the element during prestraining. In the present section, we shall explain why the L_p and the L_n curves *are not* symmetric with respect to the σ-axis. In the next section, we shall give the reason why the analogous σ/τ yield curves obtained by considering the overall response of the specimen *are* symmetric with respect to that axis.

There are two ways to show that uniaxial plastic prestraining produces local σ/τ yield curves that are unsymmetrical with respect to the σ-axis. The first is a based on physical arguments, the second on geometrical ones.

8.2.1 Physical explanation

Let us refer to any two-dimensional infinitesimal element, as thick as the bar. In Fig. 8.2.1a, such an element is represented as a square and denoted by *abcd*. The same figure shows a possible pattern for the plastic shearing flow lines produced in the specimen by a uniaxial prestraining process. It should be appreciated that, in a material element, this process can activate either a p-mode or an n-mode plastic prestraining, due to the infinitesimal dimensions of the element itself. As a consequence, the subsequent application of shear stress to the sides of the element is bound to produce different yield limits depending on the sign of the applied shear. Suppose for instance that due to the prestraining process, the plastic shearing lines of the element happen to be directed as in Fig. 8.2.1b. If we apply a positive pure shear stress to that element, the tensile principal stress will be *normal* to the above lines. By contrast, a negative pure shear stress will make the tensile principal stress *parallel* to the same lines. Clearly, these are two different physical situations. As such, they will in general produce different elastic limits. This will make the σ/τ yield curves of the element unsymmetrical with respect to the σ-axis (cf. e.g. curves L_p and L_n in Fig. 8.2.1c). Such a conclusion is in agreement with the predictions from eq. (6.4.12).

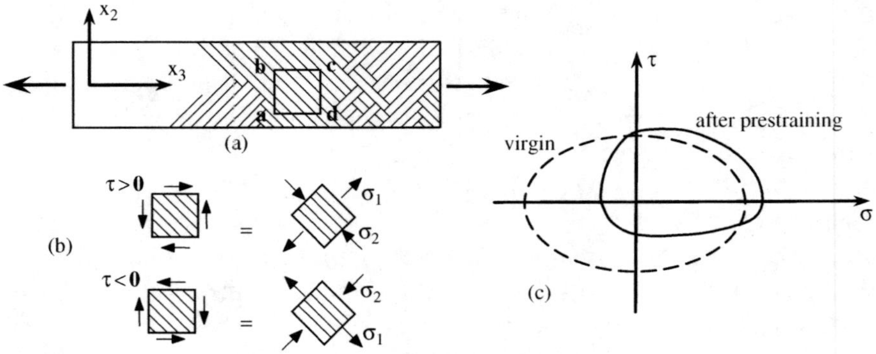

Fig. 8.2.1 (a) Possible pattern of p- and n-lines in the bar, resulting from uniaxial stress prestraining. (b) Inverting the sign of the shear stresses subsequently applied to a bar element switches the directions of the ensuing principal stresses from normal to parallel to the plastic shear lines. (c) This makes the yield limit of the element depend on the sign of the applied shear.

8.2.2 The geometry behind the σ/τ yield curve asymmetry

The lack of symmetry in τ of the σ/τ yield curves following uniaxial stress prestraining is consistent with the theoretical predictions we made in Ch. 6. As observed above, eq. (6.4.12) produces unsymmetrical yield curves with respect to σ-axis when applied to such a situation. The following arguments give a geometric explanation to this.

We may start by recalling that any two states of opposite pure shear share the same principal directions. In the case of the bar considered above, the shear plane coincides with plane (x_1, x_2). Two of the principal directions of such opposite states of pure shear belong to that plane and form an angle of $\pm\pi/4$ rad with the bar axis. The third principal direction is normal to them. Both states of shear have, therefore, the same principal axes (though opposite principal values). Accordingly, the yield limits relevant to both of them belong to the same **Q** cross-section of the yield surface (cf. Sect. 6.2). In the present case, tensor **Q** represents an anticlockwise rotation of $\theta_3 = \pi/4$ rad about the x_3-axis.

In order to get a graphic representation of this particular **Q** cross-section of the yield surface, let us refer to the geometric representation we introduced earlier in Fig. 6.2.1. The plane $\psi_{dev} = \kappa^2$ of that figure is shown in Fig. 8.2.2. The circle in dashed lines represents the **Q** cross-sections of the yield surface of the virgin material. These cross-sections are all represented by

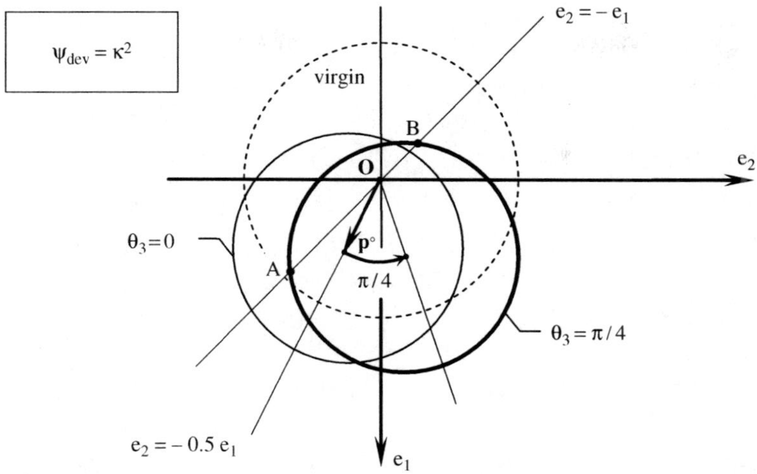

Fig. 8.2.2 **Q** cross-sections of the yield surface after uniaxial plastic prestraining. Points A and B are the new yield limits in pure shear after prestraining.

the same circle, because the yield function of the virgin material is supposed to be isotropic.

Subsequent uniaxial plastic prestraining destroys the isotropy of the yield surface. As a consequence, different **Q** cross-sections of the new yield surface will produce different circles. The particular **Q** cross-section corresponding to the states of (elastic) strain whose principal directions coincide with the coordinate axes x_1, x_2 and x_3 can be drawn immediately. It is the circle labelled $\theta_3 = 0$ in Fig. 8.2.2. Its centre lies on line $e_2 = -0.5e_1$, since in the case of uniaxial plastic prestrain, the components of vector $\mathbf{p}°$ are given by eq. (7.4.3). The diameter of this circle will generally be different from that of the circle that represents the cross-sections of the virgin material, because it depends on the value of constant b appearing in eq. (6.2.1). The actual value of this diameter as well as that of $\mathbf{p}°$ is, however, irrelevant as far as the present arguments are concerned.

Starting from cross-section $\theta_3 = 0$, we can easily obtain the **Q** cross-section that contains the two states of opposite pure shear we considered above. We simply have to rotate that cross-section by an angle of $\theta_3 = \pi/4$ rad about the ψ_{dev}-axis (normal to the plane of the picture). The resulting **Q** cross-section is labelled $\theta_3 = \pi/4$ in the same picture. All states of pure shear strain that lie on the plane of the bar belong to line $e_2 = -e_1$. This line intercepts cross-section $\theta_3 = \pi/4$ at points A and B. These are the pure shear yield limits of the bar after the considered plastic prestraining process. The picture makes it apparent that these points are not symmetric with respect to the origin of the coordinate axes. This spells different yield limits for positive and negative shear. The present theory, therefore, leads again to the conclusion that the considered axial prestraining is bound to produce non-symmetric σ/τ yield curves about the σ-axis.

8.3 OVERALL SPECIMEN RESPONSE: N/T TESTS PRODUCE SYMMETRIC σ/τ YIELD LIMITS IN SHEAR AFTER UNIAXIAL PRESTRAINING

A standard N/T test always generates a uniform state of stress and strain in the specimen at any time during the testing process. This makes the determination of the σ/τ yield curve straightforward, once forces and displacements at the ends of the specimen are measured. As already observed, however, after uniaxial plastic prestraining, the yield curves that can be determined from the test do not entirely coincide with the local σ/τ yield curves we have been considering in the previous section, be they the L_p or the L_n ones. First of all, the experimental curves thus obtained *are* symmetric with respect to the σ-axis, while generally speaking, the L_p and L_n curves are not. Secondly, a closer look at the experimental results reveals that the yield curves obtained from the tests coincide with the L_p (or L_n) curve for $\tau > 0$ and with the L_n (or L_p) curve for $\tau < 0$.

We shall now give a rationale of these apparent discrepancies between local and overall yield curves. It should be borne in mind, however, that the problem only concerns the σ/τ yield curves that the bar exhibits after uniaxial stress prestraining starting from virgin conditions. Only in this case can two different plastic prestraining modes coalesce in the bar when a given axial displacement is imposed to its boundary. If the bar is also prestrained – or was prestrained – in shear, the plastic motion of its points during the test process is unique and the coalescence of different prestraining modes is therefore ruled out. Though confined to uniaxial prestraining, the problem is nonetheless essential to a correct interpretation of the experimental results.

The arguments that follow apply to the case in which both the p-mode and n-mode motions are activated during uniaxial prestraining. This is expected to be the usual case in a finite dimension specimen of polycrystalline material, as both modes have the same likelihood of occurring. Of course, different modes will be activated at different points of the bar. Under the usual experimental conditions, however, both modes are expected to be evenly scattered throughout the specimen and, moreover, be active during the same uniaxial prestraining process. In other words, we are excluding the very special – still possible – occurrence of single mode prestraining throughout the bar. In that case, the overall specimen response at yield would of course coincide with the local curve (L_p or L_n) relevant to the particular mode that was activated by the prestraining process and the arguments that follow would not apply.

The coalescence of two different prestraining modes during the same prestraining process makes the bar non-homogeneous as far as the σ/τ yield curve of its points is concerned. In other words, the elastic limits at a point of the bar will be given by curve L_p or curve L_n, depending on which prestraining mode was last activated at that point during the prestraining process. We therefore face the task of determining which of these curves will control the elastic limit of the bar under the combined effect of N and T. The answer will be different depending on the sign of N and on the direction of the last axial plastic strain $\Delta\varepsilon_1^p$ during prestraining. We shall accordingly distinguish among the four different combinations of N and $\Delta\varepsilon_1^p$ that are specified in Fig. 8.3.1.

To begin with, let us confine our attention to case (a) of Fig. 8.3.1.

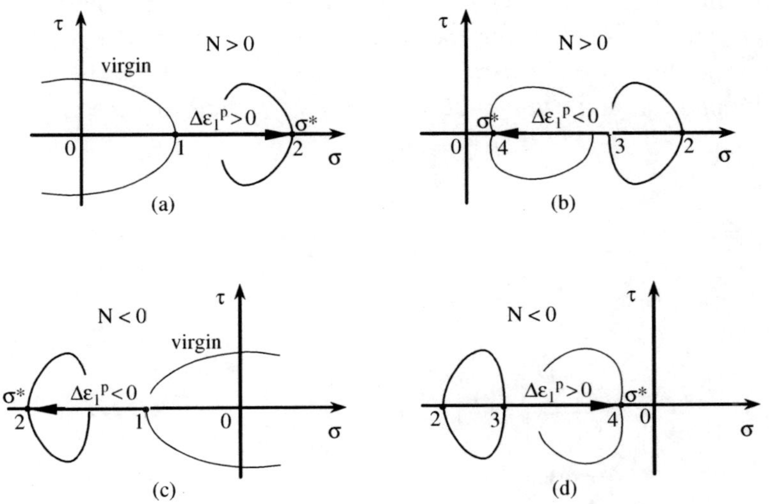

Fig. 8.3.1 Distinguishing from among different final conditions after uniaxial plastic prestraining. (a) Prestraining in tension (N > 0) ending with an axial plastic elongation ($\Delta\varepsilon_1^p > 0$). (b) Ditto, ending with an axial plastic shortening. (c) Prestraining in compression (N < 0) ending with axial plastic shortening ($\Delta\varepsilon_1^p > 0$). (d) Ditto, ending with axial plastic elongation.

8.3.1 Case (a): N > 0 and $\Delta\varepsilon_1^p > 0$

The specimen is acted upon by a tensile force (N > 0) and the last part of the prestraining process produces axial plastic elongation in the bar ($\Delta\varepsilon_1^p >$ 0). Let σ^* be the last value assumed by the axial stress of the bar at the end of the prestraining process. Figure 8.3.2a shows the directions of the plastic shearing flows in the bar in the last phase of the prestraining process. Since $\Delta\varepsilon_1^p > 0$, we must argue that the last plastic flow along the p-lines took place upwards, whereas the analogous flow along the n-lines took place downwards. (We are assuming here that the upward direction is that of the arrow of the x_2-axis.) Reference to Fig. 8.3.2b should make it clear that subsequent application of a positive shearing force T (and hence a positive shear stress τ) will favour further plastic shearing along the p-lines. The same shearing force, moreover, will oppose further plastic flow along the n-lines.

Suppose then that we keep axial force N (and thus axial stress σ) fixed at any value within the newly acquired elastic range of the material and that we gradually increase the value of T (and hence of τ) starting from T = 0. In view of the above remarks, we expect that, as T is appropriately increased, the plastic flow will continue along the p-lines. No further plastic flow is expected to take place along the n-lines, simply because the prestraining process produced a downward flow along these lines and a positive value of T will oppose it. In these conditions, we can conclude that the elastic limit

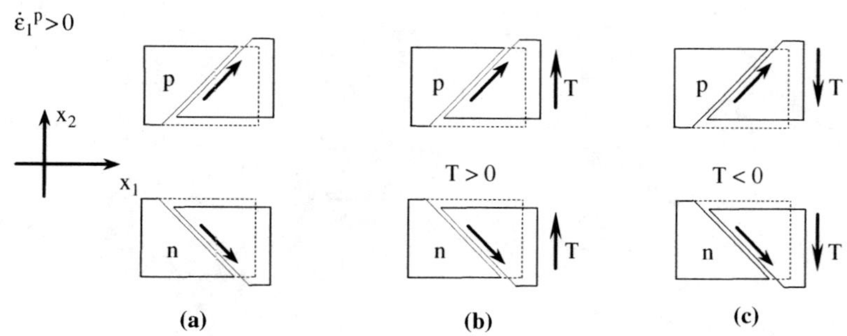

Fig. 8.3.2 (a) Plastic axial elongation ($\dot{\varepsilon}_1^p > 0$) produces opposite plastic flows in the x_2-direction depending on whether it occurs in the p- or the n-mode. (b) Subsequent application of a positive shearing force T favours further plastic flow along the p-lines while hindering plastic flow along the n-lines. (c) The converse holds true if T is negative.

of the bar for $\tau > 0$ will be given by the σ/τ yield curve that controls plastic flow along the p-lines, namely the L_p curve.

In drawing the above conclusion, we implicitly excluded that for $N > 0$ and $T > 0$ upward plastic flow can take place along the n-lines. If such a flow began before the analogous flow along the p-lines, the yield limit of the bar for $\tau > 0$ would be given by curve L_n rather than L_p. It is not difficult to realize, however, that in such a case the material would be unstable in Drucker's sense. This can be seen more clearly by referring to Fig. 8.3.3. For $N > 0$ and $T > 0$, any plastic flow along the n-lines would invariably make T and N perform opposite work. Any portion of the force that performs negative work could then be taken as Drucker's 'external agency'. The latter could be applied to the specimen while the remaining forces are kept acting on the specimen. If plastic flow occurred along the n-line, a cycle of application and removal of such 'external agency' would make it perform a negative network on the specimen, which is not possible if the material is stable in Drucker's sense. In a stable material in Drucker's sense, therefore, plastic flow under the considered conditions must occur along the shorter p-lines. This is tantamount to saying that for $\sigma > 0$ and $\tau > 0$, the σ/τ yield curve that controls the yielding of the specimen is the L_p one.

It will be noted that in the above arguments and in Fig. 8.3.3, we neglected the shear forces acting on the long sides of the specimen. This is quite correct since the bar is supposed to be equivalent to a thin-walled tube in tension/torsion. In such a case, the shear forces on the long side of the bar cannot be applied independently of T. They are the internal stress resulting from it. As such, they cannot be regarded as external forces and, therefore, do not play any role in the above arguments concerning the 'external agency'.

It remains to be considered what happens if $T < 0$ (or equivalently $\tau < 0$). The situation relevant to this case is shown in Fig. 8.3.2c. A glance at this figure should be enough to conclude that a negative value of T opposes further upward plastic flow along the p-lines, while favouring downward plastic flow along the n-lines. The same chain of arguments as the ones

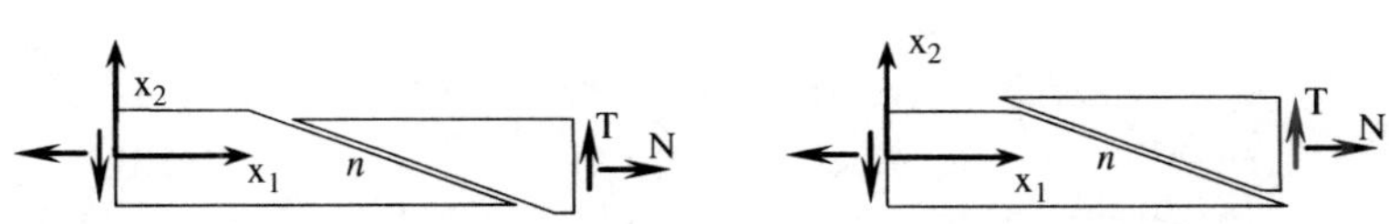

Fig. 8.3.3 Either way, plastic shearing flow along n-lines makes positive valued external forces T and N perform work of opposite sign.

pursued above will then lead us to infer that for $\sigma > 0$ and $\tau < 0$, the curve that controls the yielding of the bar is the L_n one.

We can conclude, therefore, that after a tensile uniaxial stress prestraining leading to $\Delta\varepsilon_1^p > 0$, the σ/τ yield curve of the bar for $\sigma > 0$ will be given by the following curves: L_p for $\tau > 0$ and L_n for $\tau < 0$.

A similar analysis can be pursued for the other cases considered in Fig. 8.3.1. The conclusions that can thus be reached are reported below. Stable material behaviour in Drucker's sense is assumed throughout.

8.3.2 Case (b): N > 0 and $\Delta\varepsilon_1^p < 0$

This case is illustrated in Fig. 8.3.1b. For $\sigma > 0$, the σ/τ yield curve of the bar after prestraining turns out to be given by the following curves: L_n for $\tau > 0$ and L_p for $\tau < 0$.

The further two cases to be considered below differ from the previous ones in that they concern plastic prestraining under uniaxial stress of compression. Reference to Fig. 8.3.4 rather than Fig. 8.3.2 should then be made to assess the influence of the sign of T on the ensuing plastic flow. Likewise, when applying Drucker's stability considerations to rule out unstable material response, Fig. 8.3.3 should be properly modified to represent plastic flow along the p-lines rather than the n-lines.

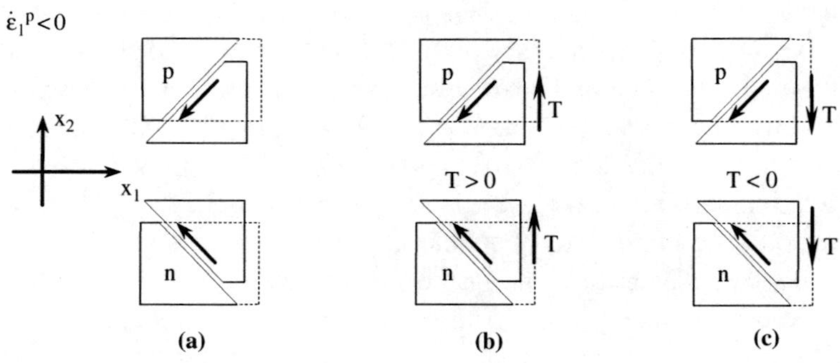

Fig. 8.3.4 (a) Similar to Fig. 8.3.2 but referring to plastic axial shortening ($\dot{\varepsilon}_1^p < 0$): opposite plastic flows are produced in the direction of axis x_2 depending on whether they occur in the p- or in the n-mode. (b) Subsequent application of a positive shearing force T, opposes further plastic flow along the p-lines while favouring plastic flow along the n-lines. (c) The converse holds true if T is negative.

8.3.3 Case (c): N < 0 and $\Delta\varepsilon_1^p < 0$

This case is illustrated in Fig. 8.3.1c. For $\sigma < 0$, the σ/τ yield curve of the bar resulting after prestraining is given by: L_n for $\tau > 0$ and L_p for $\tau < 0$.

8.3.4 Case (d): N < 0 and $\Delta\varepsilon_1^p > 0$

This is the case illustrated in Fig. 8.3.1d. After prestraining, the σ/τ yield curve of the bar for $\sigma < 0$ turns out to be given by: L_p for $\tau > 0$ and L_n for $\tau < 0$.

The results relevant to cases (a) to (d) are tabulated in Table 8.3.1. This is useful in finding out which σ/τ yield curve should apply after any uniaxial stress prestraining starting from virgin conditions. The analytical expressions of curves L_p and L_n can be obtained from eq. (6.4.12). Since they mirror each other with respect to axis σ, the expression of one of them is obtained by changing τ into $-\tau$ in the expression of the other. In view of eq. (6.4.12), this simply means changing θ_3 into $-\theta_3$. Which sign should be used for which curve depends on the adopted convention for the positive value of τ. If the convention of Figs 6.4.1c and 8.2.1b is adopted, then curve L_p will be the one that corresponds to the value of θ_3 given by eqs (6.4.3)–(6.4.5).

An example of how curves L_p and L_n join together to define the σ/τ yield curves after uniaxial stress prestraining is given in Fig. 8.3.5. The subsequent yield curves reported there refer to the particular prestraining process under tensile axial stress shown in the upper right-hand corner of the figure.

Table 8.3.1 L_p and L_n curves making up the σ/τ yield curve of flat bars and thin-walled tubes after uniaxial stress prestraining starting from the virgin material

	$\sigma > 0$		$\sigma < 0$	
	$\tau > 0$	$\tau < 0$	$\tau > 0$	$\tau < 0$
$\Delta\varepsilon_1^p > 0$	L_p	L_n	L_n	L_p
$\Delta\varepsilon_1^p < 0$	L_n	L_p	L_p	L_n

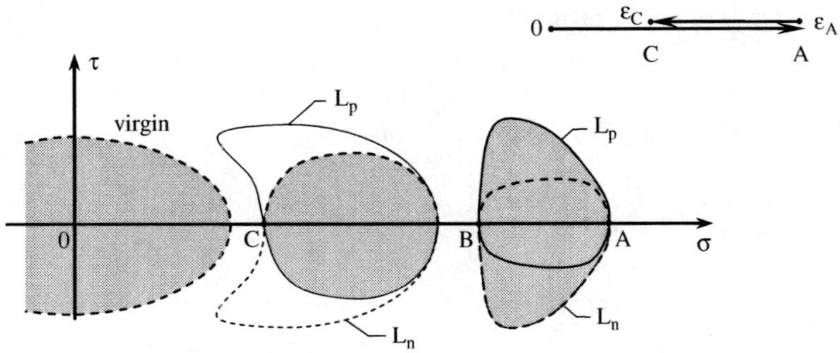

Fig. 8.3.5 Representation of symmetric σ/τ elastic regions (shadowed) showing the overall specimen response from an N/T test of a flat bar or a thin-walled tube. The two subsequent regions that follow the virgin elastic region refer to two different stages of a uniaxial plastic prestraining process (represented in the top right-hand corner of the picture) running in two subsequent opposite directions.

8.4 MATERIALS THAT ARE NOT STABLE IN DRUCKER'S SENSE

Of course, there are real materials that are not stable in Drucker's sense or lose such a stability after cold work. The negative slope of the uniaxial stress–strain curve that many materials exhibit after a large plastic strain provides the most common instance of instability following cold work. Similar instances for two- or three-dimensional stress processes are not equally well known. Simple common sense suggests, however, that in these cases too, loss of stability after a large enough plastic strain should be the rule for many ductile materials, rather than the exception.

When a material behaves as a non-stable material, the above arguments about the part of the L_p or L_n curve that should bound the σ/τ elastic region no longer apply. Neither part of these curves can *a priori* be excluded, stability arguments being ruled out. In such a situation, the yield limit of the material will lie on whichever curve turns out to be the inner one, respectively for $\tau > 0$ and for $\tau < 0$. This conclusion simply results from the fact that the inner curve will be the first one to be crossed by any process that leaves the elastic region.

Whilst on the subject of non-stable materials, it may be worth remembering that the convexity of the yield surface is a necessary condition for Drucker's stability, but by no means a sufficient one. For instance, due to cold work, an initially convex yield surface may well shrink while remaining convex. Such behaviour can be considered as a generalization of what is usually referred to as *softening*, indicating the appearance of a negative slope in the uniaxial stress–strain diagram, mainly before rupture. This is perhaps the most obvious instance of Drucker's instability – in fact the one that motivated the formulation of Drucker's postulate itself. A shrinking yield surface will reproduce the same phenomenon for a more general state of stress, quite apart from whether the yield surface is convex or not. The phenomenon occurs quite frequently in the one-dimensional case after large enough strains. It seems reasonable to expect, therefore, that a similar loss of material stability should be the rule – rather than the exception – under general states of stress for sufficiently large strains.

8.5 THE CASE OF THICK SPECIMENS

The results of this chapter refer to thin-walled circular tubes and flat bars. These are the shapes of choice to test a material under the combined action of axial and shearing forces. The same results may not apply to specimens of different cross-sections. Take a thick bar or a solid cylindrical specimen, for instance. When acted upon by an axial force, a specimen of this shape is not in a plane state of strain. In terms of decomposition (8.1.4), this means that component $\boldsymbol{\varepsilon}''^{\mathrm{p}}$ cannot be ignored. As a consequence, plastic flow – though isochoric – need not be confined to just two slip directions, nor need it be a pure shearing flow. This eliminates the non-uniqueness of the motion of the specimen points during uniaxial plastic prestraining that we detected in the case of thin bars. The consequence is that the analysis of the previous sections does not apply.

Of course, it may not be easy, or even feasible, to test thick specimens under combined axial and shear stress while keeping their state of stress and strain uniform. If that could be done without losing uniqueness in the displacement field, then their overall σ/τ yield curves would not be different from the local yield curve resulting from eq. (6.4.12) for the appropriate values of $\boldsymbol{\varepsilon}^{\circ}$ and b.

9

Theory versus Experiment

In order to predict the plastic yield limit of a work-hardened material, we need to know the details of its previous plastic deformation process. This is almost a truism and also common sense. Yet, a closer look at the wealth of experimental papers on subsequent yielding reveals that most of them fail to give enough details on the plastic strain suffered by the specimen during the experiments. More often than not, this strain is not even considered.

The reason for this may lie in the ingrained custom of regarding the yield surface in stress space. While this is a perfectly admissible standpoint, it takes no account of the fact that, by producing microscopic distortions, plastic deformation makes the stress-free state of the material change both in strain and in strain energy – a phenomenon that is fundamental to a complete description of subsequent yielding. In order to obtain experimental information concerning these changes, one has to leave the stress space and consider the yield surface as a function both of elastic deformation (or stress) and of previous plastic deformation; which clearly cannot be done if the experimental data concerning the yield surface are given in stress space, simply as functions of stress and stress path. Failure to recognize this elementary point means that today, after more than half a century of experiments on subsequent yielding, we still lack enough experimental information about the post-elastic behaviour of even the most common structural materials.

The experiments on aluminium by Phillips and Lu [9] and by Phillips and Tang [10] are among the few exceptions. They give a fairly complete record of the elastic-plastic deformation of the material in combined tension/torsion tests at various stages of the process. We shall deal with them in Sects 9.1–9.5.

In the remaining sections, we shall discuss some other often quoted experimental papers by different authors. These papers do not give enough information about plastic strain for us to make precise theoretical predictions. For this reason, we shall be content to show that the qualitative features of the results they present are indeed consistent with our theoretical predictions. A complete review of the experimental literature on the subject is, however, outside the scope of this book.

9.1 PHILLIPS AND LU'S TENSION/TORSION TESTS ON PURE ALUMINIUM

In a well-known paper dating back to 1984, Phillips and Lu [9] presented their results from a set of combined tension/torsion experiments on thin-walled tubes of commercially pure aluminium. Actually, these authors performed two series of experiments. The first series concerned stress-controlled tests and the second strain-controlled ones. In what follows, we shall confine our attention to the latter series of experiments, since paper [9] gives all the strain components that are needed to determine the elastic and the plastic strain tensors of the specimen at the main points of the deformation path considered. We shall thus be able to apply the work-hardening rule we presented in Ch. 7, and hence compare our theoretical predictions with the experimental findings of Phillips and Lu. In doing so, we shall assume that factors r and γ have the particular form we specified in Sect. 7.3.

The strain path relevant to the considered series of tests is illustrated in Fig. 9.1.1. In this figure, the purely elastic parts of the strain path are represented in dashed lines, while solid lines indicate the elastic-plastic parts of the path. Phillips and Lu determined the initial virgin σ/τ yield curve of the material and called it **I**. They also determined the subsequent yield curves as the specimen reached points C, G and K of the strain path. Following their original notation, we shall refer to them as the yield curves **II**, **III** and **IV**, respectively.

Table 9.1.1 collects together the values of total strain components ε_{11} and ε_{12} and stress components σ and τ (coinciding with σ_{11} and σ_{12}, respectively) attained by the specimen at the main points of the strain path. The values of these quantities at points C, D, E, G, H, J and K are explicitly given in Phillips and Lu's paper. Those at points L_0 and L_1 have been evaluated here from the diagrams contained in that paper, while the values relevant to points L_2 and L_3 originate from our theoretical prediction of the yield

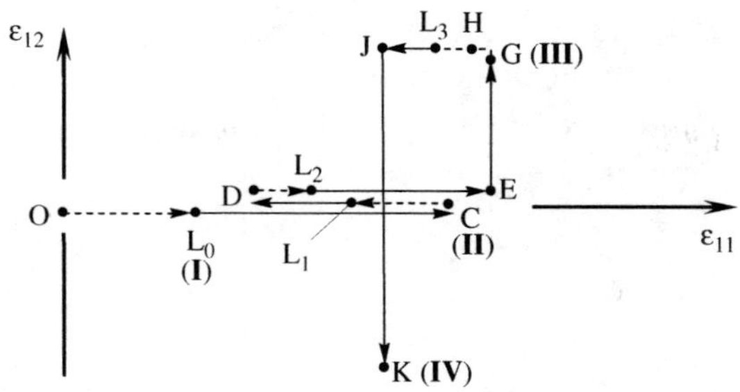

Fig. 9.1.1 Strain path of the strain-controlled tension/torsion tests made by Phillips and Lu [9]. (Capital Roman numbers refer to the experimentally determined yield surfaces at the point of the path to which they are appended.)

Table 9.1.1 Experimental values of strain and stress components at the main points of the strain path of Fig. 9.1.1, as taken from [9].

	$\varepsilon_{11} \times 10^{-6}$	$\varepsilon_{12} \times 10^{-6}$	σ (MPa)	τ (MPa)
L_0	364	0	25.50	0
C	1589	0	64.79	0
L_1	1298	0	44.43	0
D	653	0	0.98	0
L_2^*	1134	0	34.70	0
E	1781	0	65.20	0
G	1781	546	55.71	20.95
H	1763	583	51.79	22.56
L_3^*	1560	582	37.60	21.70
J	1407	582	26.20	21.20
K	1407	−857	24.70	−33.18

*Theoretically calculated values, not given in [9].

surfaces at point D (not determined by Phillips and Lu) and at point G, respectively. Here, and in what follows, reference is made to the coordinate system (x_1, x_2, x_3) that we specified in Fig. 6.4.1b.

9.2 NUMERICAL CALCULATIONS BASED ON THE EXPERIMENTAL DATA OF PHILLIPS AND LU

9.2.1 Calculating $\boldsymbol{\varepsilon}$, $\boldsymbol{\varepsilon}^e$ and $\boldsymbol{\varepsilon}^p$

The experimental data collected in Table 9.1.1 are sufficient to determine all the components of tensors $\boldsymbol{\varepsilon}$, $\boldsymbol{\varepsilon}^e$ and $\boldsymbol{\varepsilon}^p$ at the considered points of the strain path. This is actually carried out as follows. First of all, it must be recalled that in the present case the stress tensor has the form (1.3.2). Accordingly, it is completely determined by σ and τ. The same quantities also determine the elastic strain tensor:

$$\boldsymbol{\varepsilon}^{e} = \begin{bmatrix} \frac{\sigma}{E} & \frac{(1+\nu)}{E}\tau & 0 \\ \frac{(1+\nu)}{E}\tau & -\nu\frac{\sigma}{E} & 0 \\ 0 & 0 & -\nu\frac{\sigma}{E} \end{bmatrix}, \tag{9.2.1}$$

as immediately results from Hooke's law (A.16).

It must now be observed that stress components σ_{13} and σ_{23} vanish throughout the considered tension/torsion process. This means that no plastic work can be done against the homologous plastic strain components. Therefore, if the test is performed on a virgin material, we must have:

$$\varepsilon^{p}_{13} = \varepsilon^{p}_{23} = 0 \tag{9.2.2}$$

throughout the test. From eq. (9.2.1) and composition rule (1.2.2), it then follows that for the considered test:

$$\varepsilon_{13} = \varepsilon_{23} = 0. \tag{9.2.3}$$

The same composition rule (1.2.2) enables us to calculate ε^{p}_{11} and ε^{p}_{12} at the points of the strain path reported in Table 9.1.1. To do so, we simply have to subtract from the values of ε_{11} and ε_{12} of Table 9.1.1 the values of ε^{e}_{11} and ε^{e}_{12}, respectively. The latter are obtained from eq. (9.2.1) for the appropriate values of σ and τ.

The remaining two components of $\boldsymbol{\varepsilon}^{p}$, namely ε^{p}_{22} and ε^{p}_{33}, can be determined by recalling that plastic deformation is isochoric and that in the present case $\varepsilon^{p}_{22} = \varepsilon^{p}_{33}$, as stems from simple symmetry arguments. It can then be concluded that:

$$\varepsilon^{p}_{22} = \varepsilon^{p}_{33} = -\frac{1}{2}\varepsilon^{p}_{11}. \tag{9.2.4}$$

All the unknown components of $\boldsymbol{\varepsilon}$ that still remain to be determined can finally be obtained as a sum of the homologous components of $\boldsymbol{\varepsilon}^{e}$ and $\boldsymbol{\varepsilon}^{p}$, as again results from eq. (1.2.2). Once the tensor $\boldsymbol{\varepsilon}$ is thus determined at the main points of the strain path, its deviatoric part $\mathbf{e}$ at the same points of the process can be obtained at once from eq. (A.6).

9.2.2 Determining $\boldsymbol{\varepsilon}^{\circ}$

As already discussed in Sect. 7.1, to predict the yield surface of a work-hardened material, we need to know the value of $\boldsymbol{\varepsilon}^{\circ}$. The latter can be determined by integrating eq. (7.2.3) along the considered deformation path. To do this, of course, the initial value of $\boldsymbol{\varepsilon}^{\circ}$ must be known ($\boldsymbol{\varepsilon}^{\circ} = \mathbf{0}$ initially, if the material is virgin at the beginning of the process). For the time being, we shall assume that both constant r and function γ are known. In Sect. 9.3 we shall see how these constitutive quantities can be determined from the experimental data we are considering.

In practice, the procedure to determine $\boldsymbol{\varepsilon}^{\circ}$ can be the following. First of all, we divide the deformation process into a finite number n of appropriately small steps. Let $t_0, t_1, \ldots, t_i, \ldots$ and t_{n-1} respectively be the time at the beginning of the 1st, 2nd, ..., (i + 1)-th, ... and n-th step. The time interval of the i-th step will therefore be $\Delta t_{(i)} = t_i - t_{i-1}$. It will be convenient to assume that the process within each time interval is either entirely elastic or entirely elastic-plastic. In other words we are excluding that in any of the above time intervals the deformation process is purely elastic in some parts of the interval and elastic-plastic in other parts of the same interval. This condition can always be met by appropriately reducing the amplitude of each time step. To indicate that a quantity, say $\boldsymbol{\varepsilon}$, is calculated at time t_k we shall write $\boldsymbol{\varepsilon}_{(k)}$. Thus $w^p_{(i)}, \gamma_{(i)}, \mathbf{e}_{(i)}, \boldsymbol{\sigma}_{(i)}, \boldsymbol{\varepsilon}^p_{(i)}$ and $\boldsymbol{\varepsilon}^{\circ}_{(i)}$ will respectively denote the values of w^p, γ, $\mathbf{e}$, $\boldsymbol{\sigma}$, $\boldsymbol{\varepsilon}^p$ and $\boldsymbol{\varepsilon}^{\circ}$ at time t_i.

At each time of the process, both stress and total strain (and hence also strain deviator $\mathbf{e}$) are known. We can therefore calculate the value of w^p at time t_i through the following approximation:

$$w^p_{(i)} = \sum_{1}^{i}{}_k \operatorname{tr}(\bar{\boldsymbol{\sigma}}_{(k)} \Delta \boldsymbol{\varepsilon}^p_{(k)}). \tag{9.2.5}$$

Here we set:

$$\bar{\boldsymbol{\sigma}}_{(k)} = \frac{1}{2}(\boldsymbol{\sigma}_{(k-1)} + \boldsymbol{\sigma}_{(k)}), \tag{9.2.6}$$

and

$$\Delta \boldsymbol{\varepsilon}^p_{(k)} = \boldsymbol{\varepsilon}^p_{(k)} - \boldsymbol{\varepsilon}^p_{(k-1)}. \tag{9.2.7}$$

The value of $\boldsymbol{\varepsilon}^{\circ}$ at time t_i can then be calculated using the formula:

$$\boldsymbol{\varepsilon}^{\circ}_{(i)} = \boldsymbol{\varepsilon}^{\circ}_{(0)} + \sum_{k=1}^{i} \gamma_{(k)} \mathbf{D}^{\mathbf{T}}_{(k)} \Delta \mathbf{e}_{(k)} \mathbf{D}_{(k)}, \tag{9.2.8}$$

which immediately results from eq. (7.2.3). Since in the purely elastic parts of the process no change in $\boldsymbol{\varepsilon}^{\circ}$ takes place, we must understand that in eq. (9.2.8) $\Delta \mathbf{e}_{(k)}$ is given by:

$$\Delta \mathbf{e}_{(k)} = \begin{cases} \mathbf{e}_{(k)} - \mathbf{e}_{(k-1)} & \text{if } \Delta \boldsymbol{\varepsilon}^{p}_{(k)} \neq \mathbf{0} \\ \mathbf{0} & \text{if } \Delta \boldsymbol{\varepsilon}^{p}_{(k)} = \mathbf{0} \end{cases} \tag{9.2.9}$$

Tensor $\boldsymbol{\varepsilon}^{\circ}_{(0)}$ appearing in eq. (9.2.8) represents the value of $\boldsymbol{\varepsilon}^{\circ}$ at the beginning of the deformation process. For the considered tests, we have $\boldsymbol{\varepsilon}^{\circ}_{(0)} = \mathbf{0}$, as the specimen is virgin at the beginning of the test. On the other hand, the orthogonal tensor $\mathbf{D}_{(k)}$ entering the same equation represents the rotation to be applied to $\Delta \mathbf{e}_{(k)}$ (or for that matter to $\gamma_{(k)} \Delta \mathbf{e}_{(k)}$) to reduce by a factor r the angles α_i that the principal axes of $\Delta \mathbf{e}_{(k)}$ form with the homologous axes of permanent elastic strain $\boldsymbol{\varepsilon}^{\circ}_{(k-1)}$. The latter is the value of $\boldsymbol{\varepsilon}^{\circ}$ at the time t_{k-1} when strain increment $\Delta \mathbf{e}_{(k)}$ is being applied. The reader is referred to Sect. 7.2 for a complete definition of angles α_i.

To calculate tensor $\mathbf{D}_{(k)}$ we can proceed as follows. To begin with, we notice that from eq. (7.2.3) or (9.2.8) it is not difficult to infer that in the considered process, axis ξ_3 (the one which parallels the x_3-axis, as illustrated in Fig. 6.4.1) of both $\boldsymbol{\varepsilon}^{\circ}$ and $\Delta \mathbf{e}$ is always normal to the stress plane. The latter is plane (x_1, x_2) according to the notation illustrated in Fig. 6.4.1, to which we shall refer consistently in what follows. At any time t_k of the process, therefore, the relative rotation of the principal triad of $\boldsymbol{\varepsilon}^{\circ}_{(k-1)}$ and that of $\Delta \mathbf{e}_{(k)}$ will be determined completely by just one single angle, say $\theta_{(k)}$, representing the rotation about axis ξ_3 that is needed to superimpose the principal triad of $\Delta \mathbf{e}_{(k)}$ onto the analogous triad of $\boldsymbol{\varepsilon}^{\circ}_{(k-1)}$. This angle is denoted by θ_3 in Fig. 6.4.1. According to eq. (B.42), rotation $\mathbf{D}_{(k)}$ will then be given by:

$$\mathbf{D}_{(k)} = \begin{bmatrix} \cos[(1-r)\theta_{(k)}] & \sin[(1-r)\theta_{(k)}] & 0 \\ -\sin[(1-r)\theta_{(k)}] & \cos[(1-r)\theta_{(k)}] & 0 \\ 0 & 0 & 1 \end{bmatrix}. \tag{9.2.10}$$

In calculating angle $\theta_{(k)}$ introduced above, care must be exercised to comply with the convention we adopted in Sect. 7.2 concerning the principal directions of $\boldsymbol{\varepsilon}^\circ$ and $\dot{\mathbf{e}}$ (whose values here are $\boldsymbol{\varepsilon}^\circ_{(k-1)}$ and $\Delta\mathbf{e}_{(k)}$, respectively). When applied to the present case, that convention implies that $\theta_{(k)}$ is the smallest angle formed by the principal axes of $\boldsymbol{\varepsilon}^\circ_{(k-1)}$ lying in plane (x_1, x_2) and the analogous principal axes of $\Delta\mathbf{e}_{(k)}$. This will clearly make $\theta_{(k)} \in (-\pi/4, \pi/4]$. The actual calculation of this angle can be pursued by first determining the angles of the principal directions of tensors $\boldsymbol{\varepsilon}^\circ_{(k-1)}$ and $\Delta\mathbf{e}_{(k)}$ with coordinate axes x_1 and x_2. These angles can be obtained by applying to the above tensors the classical formula:

$$\theta = \frac{1}{2}\,\mathrm{atan}\frac{2q_{12}}{q_{11} - q_{22}}, \qquad \theta \in \left(-\frac{\pi}{4}, \frac{\pi}{4}\right]. \tag{9.2.11}$$

Here q_{ij} stands for the ij-component of any plane tensor $\mathbf{q}$ whose plane coincides with plane (x_1, x_2), while θ is the smallest of the two angles that the principal axes of $\mathbf{q}$ lying in plane (x_1, x_2) form with the axis x_1. Once the directions of the principal axes of $\boldsymbol{\varepsilon}^\circ_{(k-1)}$ and $\Delta\mathbf{e}_{(k)}$ are thus determined, calculating the value of $\theta_{(k)}$ according to the adopted convention is a straightforward task.

9.3 EVALUATING r AND γ

Once $\boldsymbol{\varepsilon}^\circ$ is determined at a point of the strain path, it is an easy matter to calculate the values of a_i and $\mathbf{R}^\circ$ at the same point (cf. Sect. 4.6 for a definition of $\mathbf{R}^\circ$). Coefficients a_i are given by eq. (5.2.5), while tensor $\mathbf{R}^\circ$ is obtained by applying eqs (B.7)–(B.9) to $\boldsymbol{\varepsilon}^\circ$. The yield surface at the considered point of the strain path can then be expressed in the form (6.1.1), (6.1.2) or (6.1.3), as preferred. The constant b, appearing in these equations, is determined by requesting that the elastic strain at which the last plastic deformation took place should belong to the yield surface (consistency condition).

Such a procedure assumes that the values of r and γ are already known. This is the most obvious assumption, since both r and γ are constitutive quantities and, as such, their expression should be specified in advance in order to characterize the material completely. No previous knowledge about them is currently available, though – a hardly surprising situation, as these quantities are defined in this book for the first time. We can, however, exploit the experimental data provided by Phillips and Lu [9] to evaluate r and γ for the material considered in their tests. This can be done through a trial-and-error procedure, the steps of which are exemplified below with reference to point C of the considered strain path.

In order to calculate the σ/τ yield curve of the material at point C of the strain path (curve **II** in Phillips and Lu's notation), let us respectively denote by t_0 and t_1 the values of the time parameter t at points L_0 and C of the path. The value of $\boldsymbol{\varepsilon}^\circ$ at L_0 will accordingly be $\boldsymbol{\varepsilon}^\circ_{(0)}$, while $\boldsymbol{\varepsilon}^\circ_{(1)}$ will be the value of $\boldsymbol{\varepsilon}^\circ$ at C. As already observed, we have:

$$\boldsymbol{\varepsilon}^\circ_{(0)} = \mathbf{0}, \tag{9.3.1}$$

since the material is virgin at L_0. The value of $\boldsymbol{\varepsilon}^\circ$ at C will then be given by:

$$\boldsymbol{\varepsilon}^\circ_{(1)} = \mathbf{0} + \gamma_{(1)}\mathbf{D}^{\mathbf{T}}_{(1)}\Delta\mathbf{e}_{(1)}\mathbf{D}_{(1)}, \tag{9.3.2}$$

as immediately results by applying eq. (9.3.2) to the time interval $[t_0, t_1]$. The tensor $\Delta\mathbf{e}_{(1)}$ appearing in this equation is known, because the strain path is given. The values of $\gamma_{(1)}$ and $\mathbf{D}_{(1)}$, however, are presently unknown, as the constitutive equations for γ and r are yet to be determined.

To proceed further, we shall start by assigning an arbitrary tentative value to r and γ. The procedure that follows will indicate how these values should subsequently be modified to obtain the final values that best fit the experimental data. It should be observed that, when dealing with branch L_0–C of the deformation process, the value of r is not needed, nor can it be determined from the experimental data relevant to this part of the process. The reason being that the vanishing of tensor $\boldsymbol{\varepsilon}^\circ$ at the beginning of the process makes this tensor trivially coaxial with $\Delta\mathbf{e}_{(1)}$, no matter the principal directions of the latter. Accordingly, angle $\theta_{(1)}$ between $\boldsymbol{\varepsilon}^\circ_{(0)}$ and $\Delta\mathbf{e}_{(1)}$ will vanish, which implies that $\mathbf{D}_{(1)} = \mathbf{1}$, as results immediately from eq. (9.2.10) once $\theta_{(1)}$ is substituted for $\theta_{(k)}$. It should be clear, however, that when dealing with other branches of the strain path, both r and γ may need to be assigned initially.

Once tentative values for r and γ are given, we can exploit eq. (9.2.8) to calculate the value of $\boldsymbol{\varepsilon}^\circ$ at the end of the considered time interval. For process L_0–C, in particular, the value $\boldsymbol{\varepsilon}^\circ_{(1)}$ at point C can be obtained from eq. (9.3.2), once $\mathbf{D}_{(1)}$ is set equal to $\mathbf{1}$. By means of eqs (B.7)–(B.10) of Appendix B, we proceed by calculating both the principal values of $\boldsymbol{\varepsilon}^\circ_{(1)}$ and tensor $\mathbf{R}^\circ$. The latter gives the rotation of the principal directions of $\boldsymbol{\varepsilon}^\circ_{(1)}$ relative to the coordinate axes. (Again, in the considered part of the process, $\mathbf{R}^\circ$ will be equal to $\mathbf{1}$, though in general it will not be such.) By inserting the principal values of $\boldsymbol{\varepsilon}^\circ_{(1)}$ into eq. (5.2.5), we calculate the values of coefficients a_i to be introduced into eq. (6.4.12) to obtain the σ/τ yield curve at the considered point of the strain path. The constant b appearing in this equation is determined by imposing that this curve should pass through the point $(\sigma_{(1)}, \tau_{(1)})$, representing the values of σ and τ at C.

It should be observed that in the present case, tensor $\mathbf{R}^\circ$ has the simple form (6.4.10). Accordingly, it is completely determined by angle β appearing in that equation, the value of which is known once $\boldsymbol{\varepsilon}^{\circ}$ is given. On the other hand, angle θ_3 appearing in eq. (6.4.12) is the one we already introduced in eq. (6.4.9), and represents the rotation about axis x_3 needed to superimpose the coordinate triad onto the principal triad of the elastic strain tensor or, equivalently, that of the stress tensor acting on the specimen. Once $\boldsymbol{\sigma}$ is known, therefore, angle θ_3 should be considered as known.

It should not be forgotten, moreover, that in Ch. 6 as well as in many other parts of the book where no risk of confusion arose, we have written $\boldsymbol{\varepsilon}$ for $\boldsymbol{\varepsilon}^{e}$ and $\mathbf{e}$ for $\mathbf{e}^{e}$, in order to have more readable equations involving the components of these tensors. In the present chapter, however, tensors $\boldsymbol{\varepsilon}$ and $\mathbf{e}$ stand correctly for total strains, elastic strains being explicitly denoted by $\boldsymbol{\varepsilon}^{e}$ and $\mathbf{e}^{e}$ as needed.

Clearly, the limit curve obtained through the above procedure depends on the tentative values we assumed for r and γ. We can actually draw this curve and check whether it affords a good fit for the experimental points of yield surface **II**, as determined in [9]. If it does not, we can attempt a new trial by choosing a new set of values for r and γ and reapplying the above procedure to determine the new values of a_i, $\mathbf{R}^{\circ}$ and b until a satisfactory match between the resulting yield surface and the experimental one is obtained.

In general, the procedure will enable us to determine a series of pairs (r, γ) at a close distance from each other, each of which leads to a reasonably good fit with the experimental data. Of course, different pairs will produce differently shaped yield surfaces, all providing a reasonably good fit with the same experimental data. The problem is then to select the right pair.

This problem can be solved as follows. First of all we repeat the above procedure with reference to the other points of the strain path where the experimental yield curves were determined. In the present case, they are points H and K. Once a set of (r, γ) pairs is determined at each of these points, we select from each set the pair (r, γ) which has about the same value of r at each of the considered points. In doing so, we try to meet assumption (7.3.1). If this assumption can in the end be met with enough accuracy, then the common value of r of the (r, γ) pairs thus selected is the value of r we were after. It may well be, however, that none of the above (r, γ) pairs shares the same value of r at every point of the strain path. In this case, assumption (7.3.1) cannot be met and a different and more realistic assumption should be introduced.

In the case of the experiments of Phillips and Lu, we found that the value $r = 0.93$ is a fairly acceptable choice for each of the considered points of the strain path except for the last one, namely point K. At that point, the value r = 0.7 appeared to provide a better fit with the experimental yield curve. It should be noted, however, that the distance between the experimentally determined surfaces is too large to enable us to draw any definite conclusion about whether and where a different constitutive assumption, than r = constant, should be adopted and what form it should have.

In order to get a more refined representation of $\gamma(\omega \cdot w^p)$, we also performed similar calculations by introducing some additional σ/τ yield surfaces beside the experimental ones. The added surfaces were theoretically calculated at points D, E and J, by guessing their position with the help of the few available experimental data, once a first estimate of r and γ based on surfaces **II**, **III** and **IV** was obtained. More experimental data appear to be needed, however, to reach a definitive assessment of the values of r and γ for the considered material.

The values of γ that we determined turned out to be consistent with the curves in Fig. 9.3.1. With all the reservations mentioned above, Fig. 9.3.1 can therefore be regarded as providing a graphic representation of eq. (7.3.2) for the kind of commercially pure aluminium that makes up the specimens of the experiments under consideration. It should be noted, however, that the curves shown in the figure involve a certain amount of guesswork. As a matter of fact, due to the distance between the experimental yield curves,

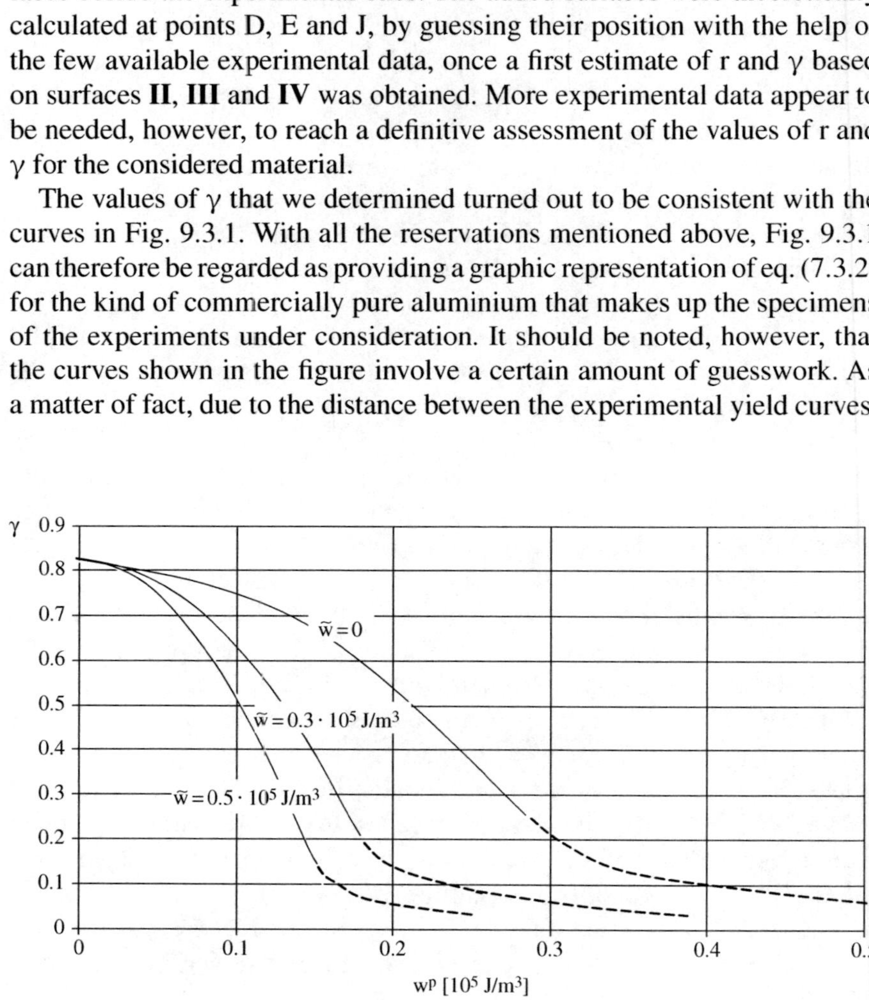

Fig. 9.3.1 Tentative curves representing the reduction factor γ for commercially pure aluminium based on the experimental data obtained by Phillips and Lu [9].

the above procedure can only give the average values of γ along the strain path joining them.

In agreement with our previous discussion on reverse loading (Sect. 7.7), in obtaining Fig. 9.3.1, we kept two separate accounts for the accumulated plastic work $\tilde{w}$ along the processes in simple tension (L_0–C and L_2–E) on the one hand and the processes in simple compression (L_1–D), on the other (cf. Fig. 9.1.1). In the remaining parts of the deformation process, the accumulated plastic work was added to the one accumulated in tension at point E, since no strict load reversal takes place from point E. It should be acknowledged, however, that if adequate experimental knowledge on the influence of the direction of the past plastic strain were available, the prediction of the subsequent yield surfaces would have to take the anisotropic past strain effect into account, as discussed in Appendix C.

From the curves of γ presented in Fig. 9.3.1, we can attempt to deduce the expression of function $\gamma = \gamma(\omega \cdot w^p)$ and the values of work-hardening factor $\omega = \omega(\tilde{w})$ we introduced through eqs (7.3.2) and (7.3.3), respectively. We shall refer to the parts of the curves represented by solid lines in Fig. 9.3.1, since the experimental data are limited to these parts. We found that these curves can be approximated by the following equation:

$$\gamma = \gamma(\omega \cdot w^p) = 0.82 - 7(\omega \cdot w^p)^2, \qquad (9.3.3)$$

where $\omega = \omega(\tilde{w})$ assumes the values that are specified in Fig. 9.3.2.

9.4 ACCURACY OF THE THEORETICAL PREDICTIONS

It remains to be seen how well the theoretical yield surfaces fit the experiment. Figure 9.4.1 shows a sequence of σ/τ yield curves as obtained from

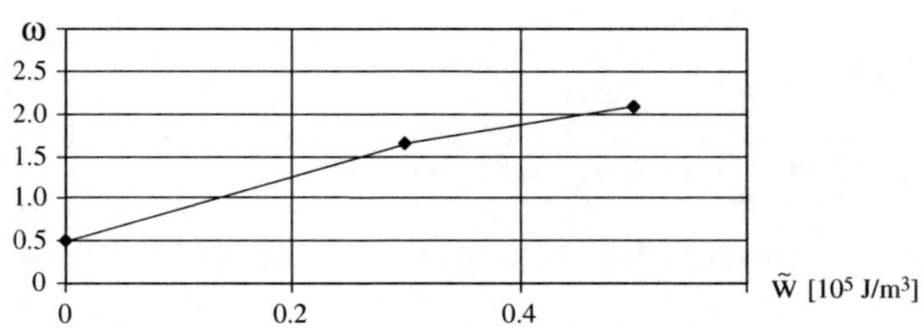

Fig. 9.3.2 Representation of function $\omega = \omega(\tilde{w})$ as resulting from the values of γ presented in Fig. 9.3.1.

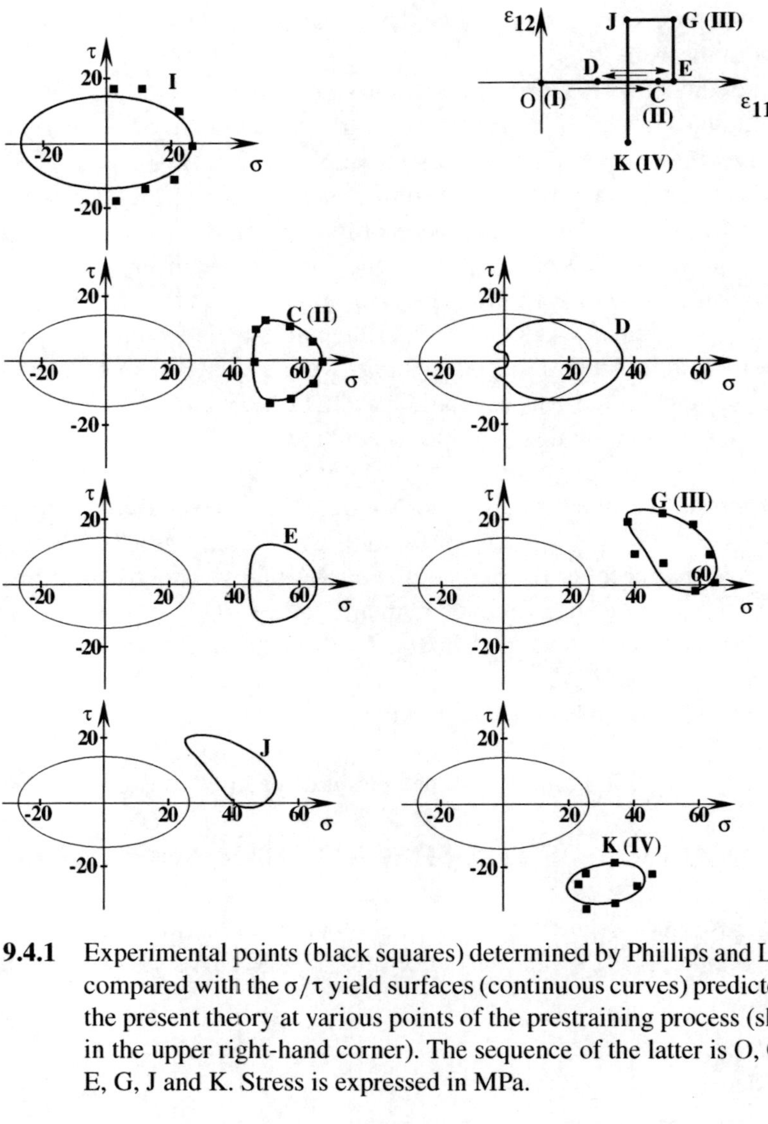

Fig. 9.4.1 Experimental points (black squares) determined by Phillips and Lu [9] compared with the σ/τ yield surfaces (continuous curves) predicted by the present theory at various points of the prestraining process (shown in the upper right-hand corner). The sequence of the latter is O, C, D, E, G, J and K. Stress is expressed in MPa.

the present theory at the main points of the strain path of Fig. 9.1.1. For easier reference, the strain path is shown at the top right corner of Fig. 9.4.1. In the same figure, the experimental points determined by Phillips and Lu [9] appear as solid squares. A comparison between the theoretical and the experimental results shows that the present theory provides an accurate model of subsequent yielding. It should be observed that some of the discrepancies between theory and experiments may well be due to the fact that the virgin material itself is not a perfect von Mises material. The experimental points of the yield surface **I**, relevant to the virgin material, appear to support this view (see Fig. 9.4.1).

Figure 9.4.1 includes also the σ/τ yield curves relevant to points D, E and G of the strain path. These are purely theoretical curves, as Phillips and Lu did not determine them experimentally. They were exploited in the previous section to increase the number of points at which function $\gamma = \gamma(\omega \cdot w^p)$ could be estimated. The picture shows that these curves are consistent with the experimentally determined ones.

It should be noted that the plastic deformation process leading to points C, D and E of the strain path is a simple tensile straining in the direction of the axis of the specimen. In agreement with what we observed in Ch. 8, the σ/τ yield curves relevant to these deformation processes were obtained by taking the appropriate parts of two different curves for $\tau > 0$ and $\tau < 0$, respectively.

One of the most distinguishing features of the present theory is that it fits well with all those dramatic changes in shape and position that the elastic range of strongly work-hardened materials can exhibit. No combination of kinematic and isotropic hardening rule can produce such a result over such a wide range of plastic deformation. It will be noticed that some of the yield curves represented in Fig. 9.4.1 are not convex. This is the case, in particular, of yield curve D and, to a lesser extent, yield curves G and J. As already noted in Sect. 1.3, the lack of convexity in the cross-sections of the complete yield surface of a work-hardened material need not indicate material instability. In the present case, the complete yield surface of the material is analytically defined by eq. (6.1.4). It is a convex surface and hence relevant to a stable material. The observed lack of convexity of some of its σ/τ cross-sections is to be ascribed to the fact that, due to yield limit anisotropy following plastic deformation, these curves are in fact non-planar cross-sections of the complete (convex) yield surface.

It is of course true that the values of r and γ that we determined above were purposely selected to obtain a good fit with the experimental data. There should be no doubt, however, that the variety of σ/τ yield curves that the theory can model and the good fit with the experiment over such a wide range of deformations cannot simply be the consequence of a particular choice of just a few scalar parameters. The latter do help in getting the right positioning and size of the subsequent yield curves. The shape of these curves, however, is determined by the yield condition of the virgin material, as expressed by eq. (3.3.1). This condition controls any subsequent yielding of the material itself.

9.5 PHILLIPS AND TANG'S TESTS

By using a direct load testing machine, Phillips and Tang [10] performed a series of experiments to determine the subsequent σ/τ yield curves from tension/torsion tests on thin-walled circular tubes of commercially pure aluminium at various temperatures. In what follows we shall confine our attention to their results at room temperature. It should be observed, however, that the results at higher temperature are consistent with those considered here. Accordingly, the discussion of this section can be extended to include yield curves at higher temperatures, provided that reference to the virgin yield surface at the considered temperature is made.

Phillips and Tang experimented on five specimens, which they labelled S-5, S-7, S-8, S-9 and S-10, respectively. Each specimen was prestrained along a different stress path in the $\sigma - \tau$ plane, and the σ/τ yield curves were determined at various points of the prestraining paths.

By repeating a similar analysis as the one presented in Sect. 9.3, we found that for all the considered specimens, the values of function $\gamma = \gamma(\omega \cdot w^p)$ that best fit the experimental results are contained between the two curves reported in Fig. 9.5.1. We assumed here that $\tilde{w} = 0$ and $\omega = 1$. The reason for this was that, with the exception of specimens S-7 and S-9, the deformation processes considered by Phillips and Tang did not involve any unloading in the elastic region followed by plastic reloading in a different direction. Accordingly, we considered each process as a single elastic-plastic

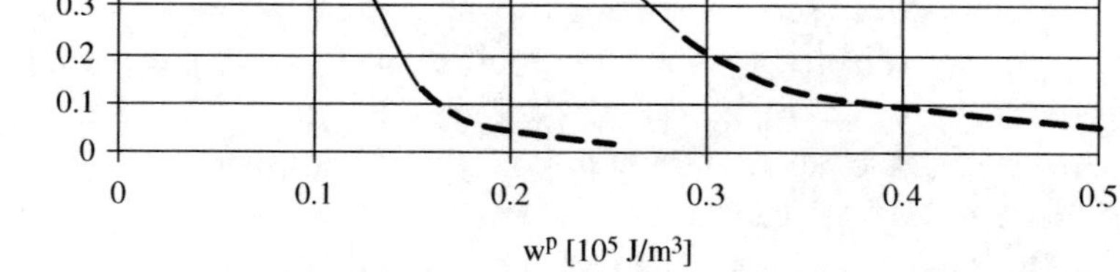

Fig. 9.5.1 Curves bounding the values of reduction factor γ as determined from the experimental data obtained by Phillips and Tang [10] on commercially pure aluminium, assuming $\tilde{w} = 0$ and $\omega = 1$ for all the tests.

deformation process starting from the virgin state. This means $\tilde{w} = 0$ and $\omega = 1$, in agreement with the hypotheses we made in Sect. 7.6. Specimens S-7 and S-9 were treated in a similar way, since the experimental data available from the paper are not sufficient to determine a more detailed dependence of γ on ω.

For some of the considered specimens, the theoretical prediction of the subsequent yield surfaces requires the knowledge of the value of the drag factor r too. These are specimens S-7, S-8, S-9 and S-10, which were subjected to elastic-plastic strain increments that turned out to be not collinear with the corresponding value of tensor $\boldsymbol{\varepsilon}^{\circ}$. For these processes, the drag factor that produced the best fit with the experiment averaged to about $r = 0.85$. More specifically, the drag factor ranged from $r = 0.75$ for specimen S-8 to $r = 1$ for specimen S-9.

Figure 9.5.2 presents the outcome of the theoretical predictions that result from the above values of γ and r. The stress path of each specimen is shown near each set of results. It should be noted that for specimen S-9, Phillips and Tang did not report the values of the plastic shear strain at the various steps of the deformation process. As a consequence some guesswork had to be made concerning these values.

In the same figure, the prestraining points at which the subsequent yield surfaces were determined are sequentially labelled from A to D, in agreement with the notation adopted in [10]. As Phillips and Tang observed, these points were not *on* the corresponding yield surface. This was probably due to the fact that the specimens were not kept long enough at the prestraining load for plastic deformation to develop fully. For this reason, in determining the theoretical curves we referred to the experimental points on the yield surface rather than the prestraining points themselves.

Even a cursory glance at the results presented in Fig. 9.5.2 shows that the correspondence of theory and experiment is generally rather good, so much so that one is tempted to ascribe the few discrepancies to errors or shortcomings of the experimental procedure. As a matter of fact, the main discrepancies are exhibited by specimens S-7 and S-9, and then only in some parts of the yield curves. They may well be a consequence of the way in which the elastic limit was determined experimentally. To determine this limit, Phillips and Tang adopted an improved backward extrapolation technique. Similar to the standard backward extrapolation technique, based on the asymptotic work-hardening line, this technique too is bound to introduce errors in the determination of the elastic limit. The magnitude of these errors depends on how sharply the stress–strain curve deviates

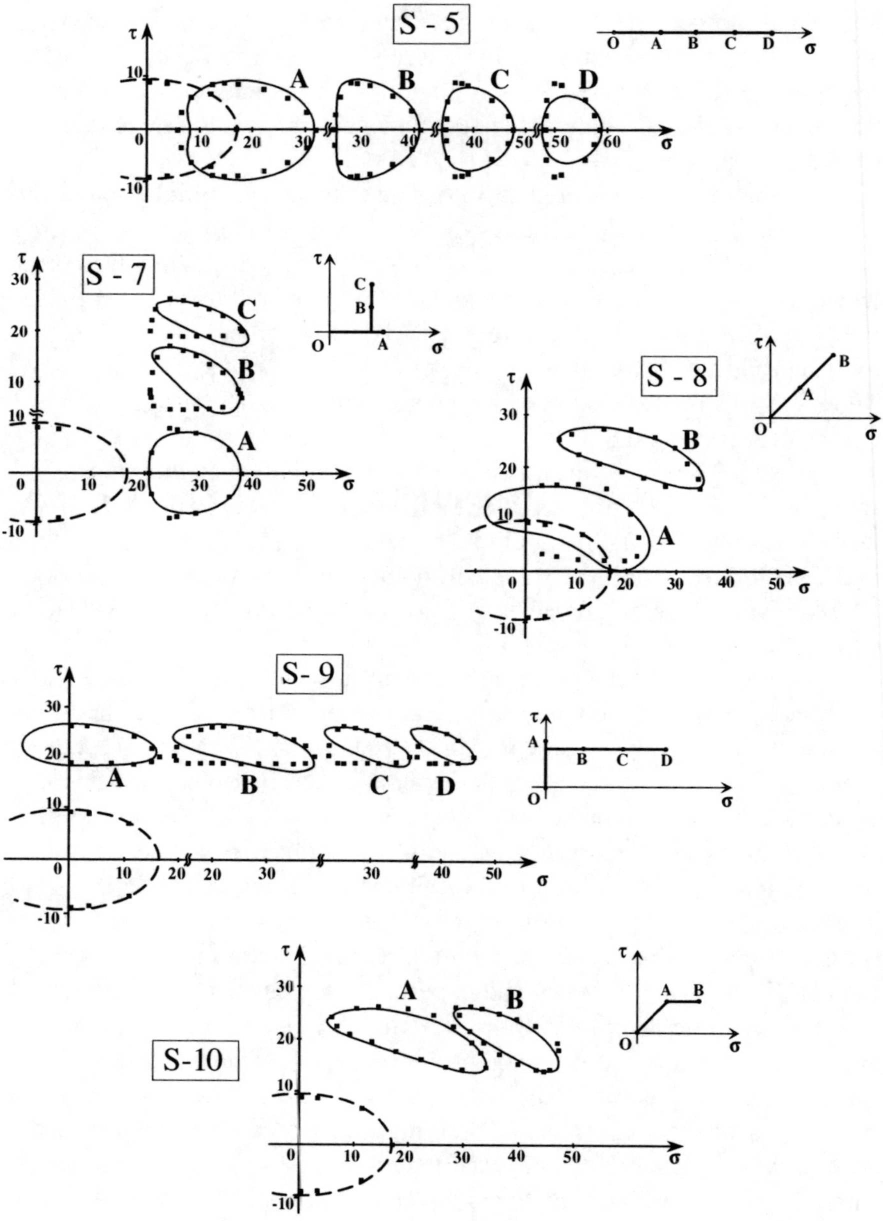

Fig. 9.5.2 Present theory predictions of the subsequent σ/τ yield curves (solid lines), compared with the experimental points determined by Phillips and Tang [10] on commercially pure aluminium at room temperature (Stress unit: MPa).

from linearity at yield. The slower the deviation, the greater is the error. As already observed in Sect. 7.8 (cf. Fig 7.7.1), inversion in the direction of plastic strain generally produces a slower elastic-to-plastic transition as plastic deformation takes place in the opposite direction. This may explain many of the discrepancies between theory and experiments appearing in Fig. 9.5.2. The phenomenon is closely related to the more general – though little studied – anisotropic past strain effect, discussed in some detail in Appendix C.

It should be observed, finally, that the two curves of γ shown in Fig. 9.5.1 and relevant to Phillips and Tang's paper, are different from curve $\gamma = \gamma(w^p)$ for $\tilde{w} = 0$ appearing in Fig. 9.3.1 and relevant to the experiments of paper [9]. The reason is that, although both papers refer to commercially pure aluminium, the material is not quite the same in the two cases. This becomes apparent when one compares the yield curves of the virgin material of the two papers, showing that the elastic limits of the virgin material are quite different in the two cases (cf. Fig. 9.4.1 and Fig. 9.5.2, respectively). It should not be forgotten, moreover, that the results discussed in Sects 9.1–9.4 refer to strain-controlled experiments, whereas those considered here are stress-controlled. In the two series of experiments, different times were allowed for the plastic strain to set. This is bound to affect the shape and size of the resulting elastic region and, hence, the values of γ too.

9.6 IVEY'S TENSION/TORSION EXPERIMENTS

In a paper dating back to 1961, Ivey [11] presented the results from a series of load-controlled tests on thin-walled tubes of various aluminium alloys. Among other experimental findings, Ivey presented three subsequent σ/τ yield curves obtained after prestraining in pure torsion a series of specimens of a silicon–aluminium alloy, referred to in his paper as 'Noral' 19S alloy. Figure 9.6.1 compares Ivey's results with the outcome of the present theory. The same figure also contains the prestraining path considered in these experiments and a plot of the function γ we calculated from Ivey's data. It should be stressed, however, that the data in Ivey's paper are not enough to determine this function accurately. Our calculation of the values of γ is based on the prestraining curve reported in Fig. 25 of [11]. From that curve, we have extracted the data relevant to the three prestraining points, needed to calculate the average value of γ in different parts of the prestraining process. The plot of function γ given in Fig. 9.6.1 should, accordingly, be regarded as tentative.

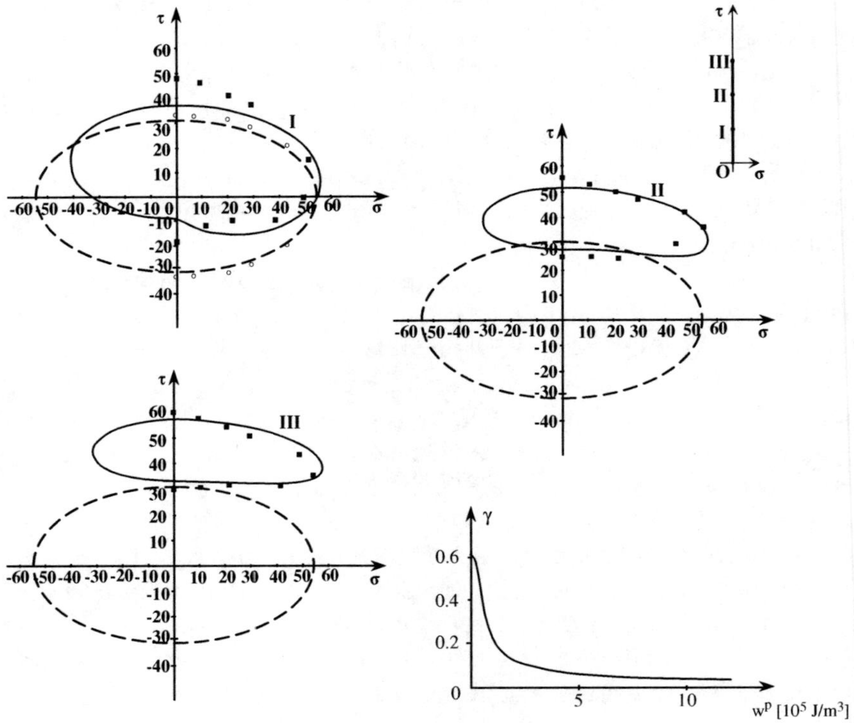

Fig. 9.6.1 Experimental points (solid squares) of Ivey's three subsequent yield curves [11], compared with predictions from the present theory (continuous curves). The dashed ellipses represent the yield curve of the virgin material. The prestraining path is shown in the top right-hand corner of the figure, while the bottom right-hand corner of the figure shows a plot of the values of the reduction factor γ resulting from the present fit.

It should be noted that plastic strain increment and tensor $\boldsymbol{\varepsilon}^{\circ}$ are coaxial throughout the considered prestraining process. This implies that the drag factor r does not enter the theoretical predictions of these yield curves nor can it be determined from such experiments.

Again, Fig. 9.6.1 shows that the present theory is capable of affording a comparatively good fit with the experiment. This is true, at least, for subsequent yield curves **II** and **III**. The experimental points of curve **I** are more at variance with the theoretical predictions. It must be said, however, that Ivey determined the elastic limit by looking for the 'first deviation from linearity' of the stress/strain plots he determined from experiment (cf.

[11, p.18]). Though conceptually correct, this method is bound to introduce subjective errors – especially so in situations where the onset of the deviation from linearity is gradual. One may wonder, therefore, whether some of the discrepancies between theory and experiment may actually be consequent to the practical difficulty in detecting this deviation.

It may also be worth observing that the experimental points relevant to the subsequent yield curve **I** show the presence of concavity in its lower part. A concavity near this part of the curve is predicted by the present theory. In Ivey's paper, however, a convex curve is drawn through the experimental points, perhaps because the possibility of non-convex yield curves in a stable material is tacitly ruled out. The matter of fact, however, may be quite different. It is widely recognized that plastic straining will generally produce anisotropy of the yield limit. As repeatedly remarked in this book, this may well produce concavities in the yield curves, simply because these curves will cease to be plane cross-sections of one single convex surface in stress space (cf. e.g. Sect. 1.3). It is not improbable, therefore, that the experimental point at the lower crossing of the yield curve **I** with the τ-axis is in fact nearer to the origin than Ivey's experimental results. A higher positioning of this point in the (σ, τ) plane would also be more consistent with the position of the two points that precede it. Errors in the experimental evaluation of the first deviation from linearity may be the true reason for the discrepancy between theory and experiment in this part of yield curve **I**.

9.7 REMNANTS OF A PAST DRAMA FROM THE EXPERIMENTAL RESULTS OF NAGHDI, ESSENBURG AND KOFF

In his paper, Ivey commented on similar experiments by Naghdi *et al.* [12] on specimens of a 24S-T-4 aluminium alloy. These authors had determined two subsequent yield curves in the σ/τ plane after plastic prestraining their specimens in pure torsion. Ivey noted that their results were at variance with his own in that they showed a less pronounced Baushingher effect and some alleged evidence of a corner at the positive crossing of the second yield curve with the τ-axis. He went on to suggest that such discrepancies were a consequence of the method followed in [12] to determine the first yield curve. He also observed that such a method was bound to influence the results relevant to the second yield curve in that it deformed the specimen far beyond the elastic domain of the first yield curve – in fact, even beyond the yield curve resulting from the subsequent prestraining.

Before assessing the results of Naghdi *et al.* in the light of the present theory, it should be observed that the mechanical properties of the particular aluminium alloy they considered were quite different from those of the material considered by Ivey. This is true, at least, if we are to judge the mechanical properties of a material from its yield stress of the virgin state. In the case of Naghdi *et al.* this stress is unusually high for an aluminium alloy, being more than seven times greater than that of Ivey's material. It should not come as a surprise, therefore, if the two materials responded differently under the same test. Yet both materials are von Mises materials and the present theory should be capable of predicting the post-elastic behaviour of both of them, no matter their particular mechanical properties. In the case of the experiments by Naghdi *et al.*, however, the theoretical prediction of the subsequent yield surfaces of the material is not as straightforward as it might at the first sight appear, as the comments below will show.

Figure 9.7.1 presents the experimental points determined by Naghdi *et al.* and the relevant theoretical curves obtained from the present theory. At this stage of the analysis, we took for granted that, as the authors claimed, the material was initially virgin and isotropic.

In order to determine curves **I** and **II**, we made the reasonable hypothesis that a pure shear plastic prestraining from the virgin state produced a tensor $\boldsymbol{\varepsilon}^\circ$ proportional to the applied shear strain, though possibly with a different coefficient of proportionality at different stages of the prestraining process. Apart from that, we did not adopted any particular work-hardening rule. The family of all the subsequent σ/τ yield curves that are compatible with the assumed hypothesis and with the yield condition of the virgin material was then obtained from eq. (6.4.12). In applying the latter equation, we observed that the present assumption on $\boldsymbol{\varepsilon}^\circ$ implied $\beta = \pi/4$ rad for every subsequent yield curve, while coefficients a_i were given by eq. (5.2.5) upon setting $\varepsilon_1^\circ = -\varepsilon_2^\circ$ and $\varepsilon_3^\circ = 0$. This makes the family of yield curves (6.4.12) depend on just two parameters, namely ε_1° and b. By selecting from this family the curves that best fitted the experimental data at hand, we finally determined the theoretical curves reported in Fig. 9.7.1.

As the figure shows, the theoretical curves thus obtained follow the overall displacement of the subsequent elastic domains with a reasonable degree of accuracy. Yet, they consistently miss the experimental data at the uppermost parts of these domains.

In trying to find an explanation for this discrepancy, we enquired whether a better fit with experiment could be obtained by removing some of the

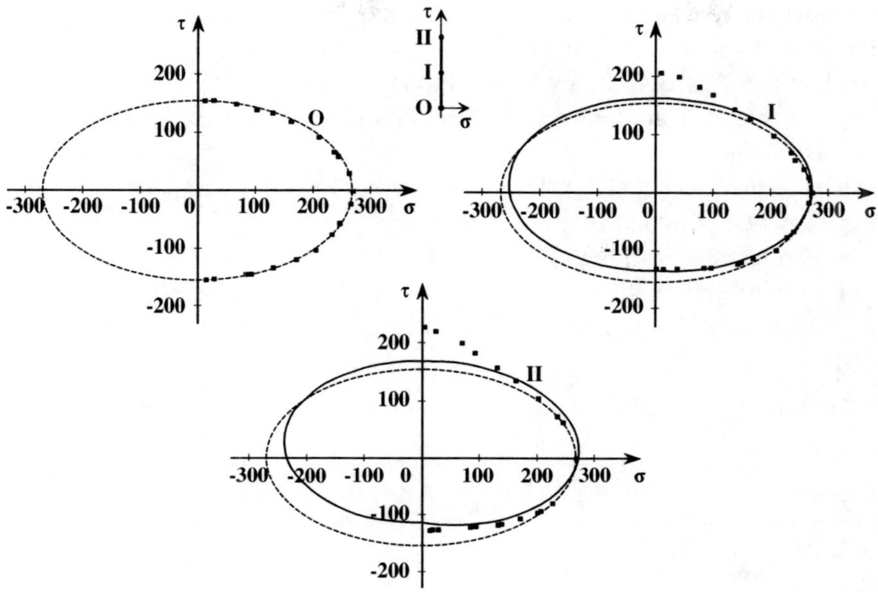

Fig. 9.7.1 Experimental points (solid squares) of the subsequent yield curves determined in [12] after pure shear prestraining (loading path shown at the top centre of the picture). The dashed ellipses represent the theoretical virgin curve (von Mises). Solid curves **I** and **II** are theoretical predictions under the assumption of initially virgin material.

restrictions we initially made on $\boldsymbol{\varepsilon}^\circ$. Accordingly, we dropped the assumption $\beta = \pi/4$, thus allowing the principal directions of $\boldsymbol{\varepsilon}^\circ$ to rotate about the ξ_3-axis (refer again to Fig. 6.4.1 for an illustration of the placement of this axis). Moreover, we left all the principal values of $\boldsymbol{\varepsilon}^\circ$ unrestricted. Also, we left the initial value of $\boldsymbol{\varepsilon}^\circ$ unspecified, thus removing the assumption that the specimen was virgin at the beginning of the test. Though rather broad, these hypotheses are not the most general ones, since they still impose that the ξ_3-axis should be normal to the plane of the applied stress – plane (x_1, x_2) in the present notation. They are, however, the minimal hypotheses under which the family of all the subsequent σ/τ yield curves is given by eq. (6.4.12). Under these conditions, this family of curves becomes much richer than before, in that it now depends on three more parameters, namely ε_1°, ε_3°, and β. Notice that we are not even assuming that $\boldsymbol{\varepsilon}^\circ$ should be traceless.

Again, by following a trial-and-error procedure, we determined the curves of this new family that best fitted the points of the two subsequent yield curves determined in [12]. Remarkably enough, in both instances the best fit was again reached for $\beta = \pi/4$ rad. Moreover, the tensor $\boldsymbol{\varepsilon}^{\circ}$ that produced the best fit with the experimental data turned out to be traceless, in spite of the fact that this condition was not imposed at the outset. The values of ε_i° and b resulting from this procedure are tabulated in Table 9.7.1 and will be discussed later.

The subsequent yield surfaces thus determined are compared with the experimental points of Naghdi *et al.* in Fig. 9.7.2.

A comparison with Fig. 9.7.1 shows that the new theoretical curves provide a much better fit with the experimental points in every part of the elastic domain. The price to be paid, however, can be read from Table 9.7.1.

Table 9.7.1 Values of ε_i° and b relevant to the theoretical yield curves reported in Fig. 9.7.2.

Curve	$\varepsilon_1^{\circ} \times 10^{-6}$	$\varepsilon_2^{\circ} \times 10^{-6}$	$\varepsilon_3^{\circ} \times 10^{-6}$	$b \times 10^5 (J/m^3)$
I	804	−2804	2000	0.42
II	922	−3236	2314	0.48

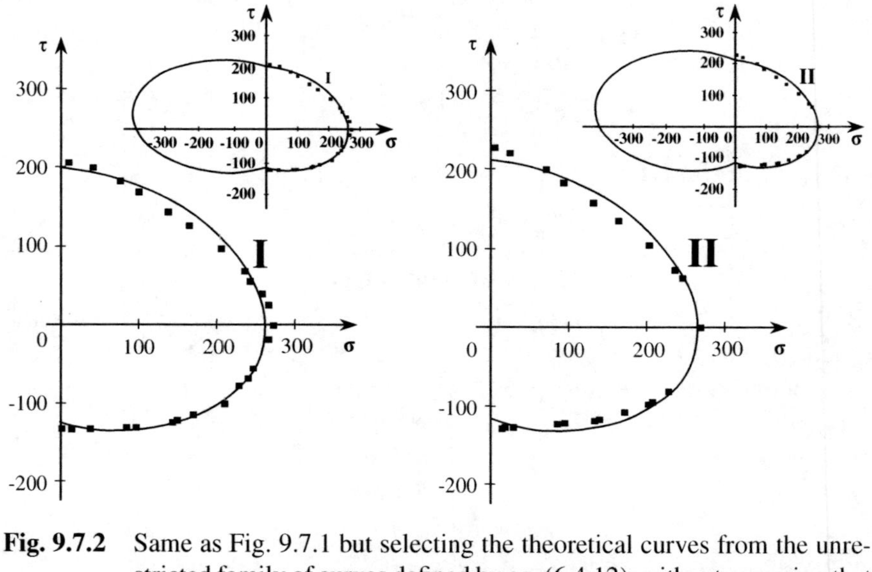

Fig. 9.7.2 Same as Fig. 9.7.1 but selecting the theoretical curves from the unrestricted family of curves defined by eq. (6.4.12), without assuming that the material was virgin at the beginning of the test. The larger diagrams refer to half-plane $\sigma > 0$, where the experimental points were actually taken. The smaller diagrams represent the complete yield curve on a smaller scale.

First of all, a non-vanishing value of ε_3° means that the material cannot be virgin at the beginning of the test. The reason for this is that a loading process in pure shear does not produce any plastic deformation normal to the shear plane. Thus, in an initially virgin material ($\varepsilon_1^\circ = \varepsilon_2^\circ = \varepsilon_3^\circ = 0$), a pure shear prestraining in shear plane (x_1, x_2) is expected to leave $\varepsilon_3^\circ = 0$, unless some very unusual work-hardening properties are assumed. Secondly, it should be noticed that Table 9.7.1 consistently gives $\varepsilon_1^\circ \neq -\varepsilon_2^\circ$. Again, this is at odds with what is expected to happen when an initially virgin material is prestrained in pure shear in the plane (x_1, x_2), since such a process produces opposite principal values of strain in that plane. Of course, the results of Table 9.7.1 are perfectly plausible if the prestraining process starts from a specimen in non-virgin conditions. In that case, any initial value for $\varepsilon_1^\circ, \varepsilon_2^\circ$ and ε_3° is in principle admissible, which however will affect the subsequent values of these quantities. Moreover, and equally important, in a non-virgin material the principal triad of $\boldsymbol{\varepsilon}^\circ$ can be initially at any angle with the axis of the specimen. This would in general make all the three principal values of $\boldsymbol{\varepsilon}^\circ$ increase as a result of plastic prestraining, even if the latter is limited to a pure shear process in the plane (x_1, x_2).

But, is there any evidence that the specimens used by Naghdi *et al.* were in fact non-virgin at the beginning of the test? The answer can be found by observing that these authors state that their specimens were taken from a round 'extruded' mother bar, whose diameter was about twice the diameter of the specimen. Although they claimed that the specimen was initially 'reasonably isotropic', it seems unlikely that the dramatic changes of shape involved in the extrusion process of the mother bar may have left it in the virgin state. As a matter of fact, its yield limit in tension, unusually high for a virgin aluminium alloy, seems to point at a strongly work-hardened material. Nor can the specimen isotropy be judged simply from the photomicrograph of its cross-section, as cursorily stated by the authors in their paper, because the symmetry of the extrusion process is bound to affect the points of the cross-section of the bar all in the same way. In the light of the results of Fig. 9.7.2, therefore, it seems plausible to conclude that the good match of the experimental data of Naghdi *et al.* with the theoretical yield curves of Fig. 9.7.2 relevant to an initially non-virgin material, indicates that the considered specimens were in fact non-virgin at the outset.

9.7.1 Some further remarks

A few more comments are in order before closing this section. First of all, it might be objected that the experimental points of the initial yield surface (namely surface O of Fig. 9.7.1) match the yield surface of a virgin von Mises material quite well. This could be interpreted as indicating that the specimens were indeed virgin at the beginning of the test, in contradiction to what was concluded above.

To counter this objection, it must be noted that the considered matching of experimental points is confined to half-plane $\sigma > 0$, since no yield point was experimentally determined in half-plane $\sigma < 0$ in paper [12]. In this limited range of values of σ, the subsequent yield curves of the material, as determined theoretically, may well exhibit a semi-elliptic shape, close to the corresponding part of the virgin ellipse, even after a comparatively strong work-hardening. It all depends on the particular prestraining process.

Take the case illustrated in the Fig. 9.7.3, for instance. In this figure, the dashed ellipse represents the yield surface of the virgin von Mises material that we already considered in Fig. 9.7.1. This curve can be obtained from eq. (6.4.12) once it is assumed that $\varepsilon_1^\circ = \varepsilon_2^\circ = \varepsilon_3^\circ = 0$ and $b = 0$ (in these conditions, the value of β is irrelevant). On the other hand, the continuous curve in the same figure represents an instance of a theoretical σ/τ yield curve obtained from eq. (6.4.12) after a comparatively strong work-hardening process. More specifically, this curve corresponds to the following values of the above parameters: $\varepsilon_1^\circ = 196 \times 10^{-6}$, $\varepsilon_2^\circ = -1372 \times 10^{-6}, \varepsilon_3^\circ = 1176 \times 10^{-6}, b = 0.20 \times 10^5 \mathrm{J/m^3}$ and $\beta = \pi/4$. Yet, for $\sigma > 0$ the new curve almost coincides with the yield curve of the virgin material.

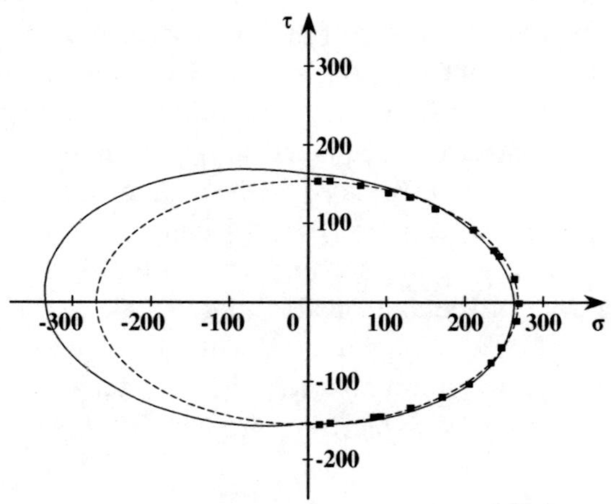

Fig. 9.7.3 An instance in which the yield curve of a von Mises material in virgin condition (dashed line) and that of the same material after strong work-hardening (continuous line) almost coincide for $\sigma > 0$.

The point to be made here is that, in order to ascertain whether the material is virgin or not, it is not sufficient to limit the experimental check to just one half-plane $\sigma > 0$ or $\sigma < 0$.

Another comment concerns the corner points that the σ/τ yield curves can exhibit at the positive crossing with the τ-axis. They are a consequence of the geometric representation of the yield curves in the (σ, τ)-plane. As discussed in Sect. 1.3 and repeatedly recalled in this chapter, corner points in the yield curves may result from the fact that these curves are in general non-planar cross-sections of the whole yield surface in a higher dimension space. In the present case, the analytical expression of the considered curves is different for $\sigma > 0$ and $\sigma < 0$, as specified by eq. (6.4.12). This indicates that they are made of two branches (relevant to $\sigma > 0$ and $\sigma < 0$, respectively) that belong to different planes in the space where the complete yield surface lies. The different slopes of the two planes may give rise to corners at the crossing with the τ-axis where the two planes meet each other, as illustrated in Fig. 1.3.1.

It may be worth mentioning, finally, that a still better fit with the experimental points considered in this section can possibly be obtained by removing the hypothesis that principal direction ξ_3 should be normal to plane (x_1, x_2). This will further widen the family of possible yield curves, which clearly cannot result in a worse fit. It should be noted that in the present case, the rotation of axis ξ_3 must be such as to leave it parallel to the (x_1, x_3)-plane (see Fig. 6.4.1), due to the axial symmetry about axis x_1 of the considered extrusion process. Observe, however, that if allowance is made for rotations of the ξ_3-axis, the equation of the family of all possible σ/τ yield curves should be calculated directly from eq. (6.1.1), because eq. (6.4.12) would no longer apply.

Epilogue

In closing this book, it may be appropriate to pause a moment and consider what we gained thus far and what we can do from here. We discovered that the subsequent yield surfaces of a von Mises material are closely related to its virgin surface (Ch. 3). We learned how to predict the family of all the possible subsequent yield surfaces that such a material can exhibit when properly strained (Ch. 6). Perhaps even more important, we spotted the variables upon which these surfaces depend (Sect. 7.1), thus setting the ground for a consistent approach to work-hardening. The problem, from that point on, becomes a constitutive one. Accordingly, different evolution rules of the work-hardening parameters are expected to apply to different von Mises materials. Some simple and yet realistic evolution rules are given in Ch. 7. These enable us to associate each yield surface of the family of all the possible yield surfaces of the material to each deformation process the material can ever undergo.

The proposed evolution rules agree reasonably well with the available experimental data (Ch. 9). They are, however, only an instance of how the present theory can be applied to predict the post-elastic behaviour of elastic-plastic materials. Further work remains to be made on this topic, which in turn calls for more experimental knowledge on the plastic response of these materials. Such knowledge is in many respects still lacking, although it is well within the reach of today's experimental capabilities.

As variously observed in the book, there are a number of topics where the results of the present analysis are actually needed. They range from Computational Mechanics to Experimental Plasticity. The former will benefit from the present theory to design software capable of producing more realistic solutions in a more general range of elastic-plastic deformation processes, thanks to more realistic work-hardening rules. Experimental Plasticity, on the other hand, is bound to take advantage of the deductive approach of this book to conceive new and more focused experiments, thanks to the backing of a firmer theoretical framework.

From the conceptual standpoint, the main questions that remain to be answered are, on the one hand, the formulation of more general and simpler work-hardening rules and, on the other, the extension of the present approach to a wider class of materials. A still more general problem that waits to be answered concerns the nature of permanent deformation and rupture, two phenomena whose irreversible character betrays their connection with Thermodynamics.

Here, we are touching the point from where all the present work originated. The only law of Physics that has a say on irreversibility is the second principle of Thermodynamics. Plastic deformation is an irreversible phenomenon, so is rupture. It is only natural to expect that this principle should be involved in such phenomena. In fact, the main ideas of this book sprang from a series of studies on the bearing of the second principle of Thermodynamics on the irreversible processes that can take place in a deforming solid. The interested reader is referred to [16] for an account of some of the relevant results. This kind of application of Thermodynamics to Solid Mechanics is, however, still in its budding stage. Whilst, the purely mechanical approach presented in this book, though not as general as it could possibly be, appears ready for practical applications.

APPENDIX A

Elastic Energy of Linear Elastic Materials

Elastic energy plays a fundamental role in the analysis of the present book. For this reason, in this Appendix, we shall recall some of the commonest expressions of it and of its volumetric and deviatoric parts.

A.1 GENERAL LINEAR ELASTIC MATERIALS

The specific elastic energy per unit volume of material will be denoted by ψ. For general linear elastic materials (whether isotropic or not), it has the well-known expression:

$$\psi = \frac{1}{2}\mathrm{tr}(\boldsymbol{\sigma}^{\mathsf{T}} \cdot \boldsymbol{\varepsilon}^{\mathrm{e}}) + \mathrm{c}^{\circ} = \frac{1}{2}\sigma_{\mathrm{ij}}\ \varepsilon^{\mathrm{e}}_{\mathrm{ij}} + \mathrm{c}^{\circ}, \tag{A.1}$$

superscript ^{T} denotes transposition. From eq. (1.2.1), we can express ψ as a function of $\boldsymbol{\varepsilon}^{\mathrm{e}}$ or $\boldsymbol{\sigma}$, according to the relations:

$$\psi = \psi(\boldsymbol{\varepsilon}^{\mathrm{e}}) = \frac{1}{2}\ \mathsf{C}_{\mathrm{ijlm}}\ \varepsilon^{\mathrm{e}}_{\mathrm{ij}}\ \varepsilon^{\mathrm{e}}_{\mathrm{lm}} + \mathrm{c}^{\circ} \tag{A.2}$$

and

$$\psi = \psi(\boldsymbol{\sigma}) = \frac{1}{2}\ \mathsf{C}^{-1}_{\mathrm{ijlm}}\ \sigma_{\mathrm{ij}}\ \sigma_{\mathrm{lm}} + \mathrm{c}^{\circ}. \tag{A.3}$$

Here, $\mathsf{C}^{-1}_{\mathrm{ijlm}}$ are the components of the compliance tensor $\mathbf{C}^{-1}$, the inverse of the elastic modulus tensor $\mathbf{C}$. Quantity c° entering the above equations is an inessential constant, representing the elastic energy of the material when all macroscopic stress $\boldsymbol{\sigma}$ is removed from it, which means $\boldsymbol{\sigma} = \mathbf{0}$. If this *stress-free state* is assumed as the reference configuration for strain, then elastic strain $\boldsymbol{\varepsilon}^e$ will also vanish in the same state. Of course, $\boldsymbol{\varepsilon}^e$ will coincide with the total strain $\boldsymbol{\varepsilon}$, until plastic deformation takes place.

It should be noted that the value of c° can be assigned arbitrarily, since energy is defined to within an arbitrary constant. For linear elastic materials, it is often assumed that no elastic energy is stored in the stress-free state. This assumption makes $c^\circ = 0$.

According to eq. (1.2.16), the stress tensor can be decomposed into the sum of a deviatoric part and a hydrostatic one according to the relation:

$$\boldsymbol{\sigma} = \mathbf{s} + \bar{\sigma}\mathbf{I}. \tag{A.4}$$

The analogous decomposition for total strain $\boldsymbol{\varepsilon}$ is:

$$\boldsymbol{\varepsilon} = \mathbf{e} + \bar{\varepsilon}\mathbf{I}. \tag{A.5}$$

Here $\mathbf{e}$ is the deviatoric part of $\boldsymbol{\varepsilon}$, also referred to as *strain deviator* and defined by:

$$\mathbf{e} = \boldsymbol{\varepsilon} - \bar{\varepsilon}\mathbf{I} \quad \text{or} \quad e_{ij} = \varepsilon_{ij} - \bar{\varepsilon}\delta_{ij}, \tag{A.6}$$

while $\bar{\varepsilon}$ is the mean strain:

$$\bar{\varepsilon} = \frac{1}{3}\mathrm{tr}(\boldsymbol{\varepsilon}) = \frac{1}{3}\,\varepsilon_{kk}. \tag{A.7}$$

Tensor $\bar{\varepsilon}\mathbf{I}$ appearing in eq. (A.6) is usually referred to as the *isotropic* (or *spherical*) component of $\boldsymbol{\varepsilon}$.

The volume change of a material element of initial volume V_o is given by $\Delta V = \mathrm{tr}(\boldsymbol{\varepsilon})\ V_o$. Since $\mathrm{tr}(\mathbf{e}) = 0$, the deformation produced by $\mathbf{e}$ does not cause any volume change. However, it will in general produce a change in the relative angles of the material lines of the element before and after deformation. For this reason, tensor $\mathbf{e}$ is also dubbed as the *distortional* component of $\boldsymbol{\varepsilon}$.

The isotropic component $\bar{\varepsilon}\mathbf{I}$ of $\boldsymbol{\varepsilon}$, on the contrary, has a different effect. It is responsible for the volume changes of the element, since $\mathrm{tr}(\boldsymbol{\varepsilon}) \equiv \mathrm{tr}(\bar{\varepsilon}\mathbf{I})$. Accordingly, it will also be referred to as the *volumetric* part of $\boldsymbol{\varepsilon}$.

A.2 DISTORTIONAL AND VOLUMETRIC COMPONENTS OF ψ (GENERAL CASE)

By applying decomposition (A.5) to elastic strain $\boldsymbol{\varepsilon}^e$, we get:

$$\boldsymbol{\varepsilon}^e = \mathbf{e}^e + \bar{\varepsilon}^e\, \mathbf{I}, \tag{A.8}$$

with the obvious meaning of the symbols. If we introduce eqs (A.4) and (A.8) into (A.1) we obtain:

$$\begin{aligned}\psi &= \frac{1}{2}\mathrm{tr}[(\mathbf{s} + \bar{\sigma}\mathbf{I})^{\mathsf{T}} \cdot (\mathbf{e}^e + \bar{\varepsilon}^e\mathbf{I})] + c^\circ = \frac{1}{2}\mathrm{tr}(\mathbf{s}^{\mathsf{T}} \cdot \mathbf{e}^e) + \frac{3}{2}\bar{\sigma}\bar{\varepsilon}^e + c^\circ \\ &= \psi_{\mathrm{dev}} + \psi_{\mathrm{vol}},\end{aligned} \tag{A.9}$$

since $\mathrm{tr}(\mathbf{s}) = \mathrm{tr}(\mathbf{e}^e) = 0$ and $\mathrm{tr}(\mathbf{I}) = 3$. The energies ψ_{dev} and ψ_{vol} appearing in this formula are defined by:

$$\psi_{\mathrm{dev}} = \frac{1}{2}\mathrm{tr}(\boldsymbol{\sigma}^{\mathsf{T}} \cdot \mathbf{e}^e) + c^\circ_{\mathrm{dev}} = \frac{1}{2}\mathrm{tr}(\mathbf{s}^{\mathsf{T}} \cdot \mathbf{e}^e) + c^\circ_{\mathrm{dev}} = \frac{1}{2}\, s_{ij}\, \mathbf{e}^e_{ij} + c^\circ_{\mathrm{dev}}, \tag{A.10}$$

and

$$\psi_{\mathrm{vol}} = \frac{1}{2}\bar{\sigma}\bar{\varepsilon}^e\, \mathrm{tr}(\mathbf{I}) + c^\circ_{\mathrm{vol}} = \frac{3}{2}\bar{\sigma}\bar{\varepsilon}^e + c^\circ_{\mathrm{vol}} = \frac{1}{6}\, \sigma_{ii}\, \varepsilon^e_{kk} + c^\circ_{\mathrm{vol}}, \tag{A.11}$$

where c°_{dev} and c°_{vol} are respectively the deviatoric and the volumetric parts of c°. Of course, $c^\circ_{\mathrm{dev}} + c^\circ_{\mathrm{vol}} = c^\circ$. Being produced by the deviatoric component of strain, ψ_{dev} is referred to as the *deviatoric* or the *distortional* component of ψ. Similarly, ψ_{vol} is the *volumetric* component of ψ, since it is due to the volumetric component of $\boldsymbol{\varepsilon}$.

A.3 ISOTROPIC LINEAR ELASTIC MATERIALS

The above expressions of ψ are valid for a general linear elastic material, as defined by Hooke's law (1.2.1). If the material is also isotropic, then the components of tensor $\mathbf{C}$ can be expressed as

$$C_{ijkl} = \lambda\delta_{ij}\delta_{kl} + \mu(\delta_{ik}\delta_{jl} + \delta_{il}\delta_{jk}). \tag{A.12}$$

Here λ and μ are the well-known Lamé constants, related to elastic modulus E, shear modulus G and Poisson's ratio ν through the relations:

$$\lambda = \frac{\nu E}{(1+\nu)(1-2\nu)} \tag{A.13}$$

and

$$\mu \equiv G = \frac{E}{2(1+\nu)}. \tag{A.14}$$

For isotropic materials, Hooke's law reduces to:

$$\sigma_{ij} = \lambda\varepsilon^e_{kk}\delta_{ij} + 2\mu\varepsilon^e_{ij}. \tag{A.15}$$

This can be inverted to give:

$$\varepsilon^e_{ij} = \frac{1}{E}[(1+\nu)\sigma_{ij} - \nu\sigma_{kk}\delta_{ij}]. \tag{A.16}$$

From these equations and from eq. (A.1) it is not difficult to express eqs (A.2) and (A.3) in the form (cf. e.g. [13, p. 85]):

$$\begin{aligned}\psi = \psi(\boldsymbol{\varepsilon}^e) &= \frac{1}{2}\lambda\varepsilon^e_{ii}\varepsilon^e_{kk} + \mu\varepsilon^e_{ij}\varepsilon^e_{ij} + c^\circ = \frac{1}{2}\lambda(\varepsilon^e_{11} + \varepsilon^e_{22} + \varepsilon^e_{33})^2 \\ &\quad + \mu({\varepsilon^e_{11}}^2 + {\varepsilon^e_{22}}^2 + {\varepsilon^e_{33}}^2 + 2{\varepsilon^e_{12}}^2 + 2{\varepsilon^e_{13}}^2 + 2{\varepsilon^e_{23}}^2) + c^\circ\end{aligned} \tag{A.17}$$

and

$$\begin{aligned}\psi = \psi(\boldsymbol{\sigma}) &= -\frac{\nu}{2E}\sigma_{ii}\sigma_{kk} + \frac{1+\nu}{2E}\sigma_{ij}\sigma_{ij} = -\frac{\nu}{2E}(\sigma_{11} + \sigma_{22} + \sigma_{33})^2 \\ &\quad + \frac{1+\nu}{2E}({\sigma_{11}}^2 + {\sigma_{22}}^2 + {\sigma_{33}}^2 + 2{\sigma_{12}}^2 + 2{\sigma_{23}}^2 + 2{\sigma_{13}}^2) + c^\circ,\end{aligned} \tag{A.18}$$

respectively.

It should now be recalled that the first and the second tensor invariants of $\boldsymbol{\varepsilon}^e$ are respectively given by:

$$I_1(\boldsymbol{\varepsilon}^e) = \mathrm{tr}(\boldsymbol{\varepsilon}^e) = \varepsilon^e_{11} + \varepsilon^e_{22} + \varepsilon^e_{33} = \varepsilon^e_1 + \varepsilon^e_2 + \varepsilon^e_3 \tag{A.19}$$

and

$$\begin{aligned}I_2(\boldsymbol{\varepsilon}^e) &= \frac{1}{2}(\varepsilon^e_{ij}\varepsilon^e_{ij} - \varepsilon^e_{ss}\varepsilon^e_{kk}) \\ &= {\varepsilon^e_{12}}^2 + {\varepsilon^e_{23}}^2 + {\varepsilon^e_{31}}^2 - (\varepsilon^e_{11}\varepsilon^e_{22} + \varepsilon^e_{11}\varepsilon^e_{33} + \varepsilon^e_{22}\varepsilon^e_{33}) \\ &= -(\varepsilon^e_1\varepsilon^e_2 + \varepsilon^e_1\varepsilon^e_3 + \varepsilon^e_2\varepsilon^e_3),\end{aligned} \tag{A.20}$$

where ε_1^e, ε_2^e and ε_3^e are the principal values of $\boldsymbol{\varepsilon}^e$. Likewise, the first and the second invariant of $\boldsymbol{\sigma}$ are given by:

$$I_1(\boldsymbol{\sigma}) = \mathrm{tr}(\boldsymbol{\sigma}) = \sigma_{11} + \sigma_{22} + \sigma_{33} = \sigma_1 + \sigma_2 + \sigma_3 \quad \text{(A.21)}$$

and

$$\begin{aligned} I_2(\boldsymbol{\sigma}) &= \frac{1}{2}(\sigma_{ij}\sigma_{ij} - \sigma_{ss}\sigma_{kk}) \\ &= \sigma_{12}{}^2 + \sigma_{23}{}^2 + \sigma_{13}{}^2 - (\sigma_{11}\sigma_{22} + \sigma_{11}\sigma_{33} + \sigma_{22}\sigma_{33}) \\ &= -(\sigma_1\sigma_2 + \sigma_1\sigma_3 + \sigma_2\sigma_3), \end{aligned} \quad \text{(A.22)}$$

where σ_1, σ_2 and σ_3 are the principal values of $\boldsymbol{\sigma}$. From eqs (A.19)–(A.22) and from eqs (A.13), (A.14), (A.17) and (A.18), the following alternative expressions for the elastic energy of isotropic linear elastic materials can be obtained:

$$\begin{aligned} \psi = \psi(\boldsymbol{\varepsilon}^e) &= \left(\frac{1}{2}\lambda + \mu\right) I_1(\boldsymbol{\varepsilon}^e)^2 + 2\mu I_2(\boldsymbol{\varepsilon}^e) + c^\circ \\ &= \psi(\varepsilon_1^e, \varepsilon_2^e, \varepsilon_3^e) = \frac{1}{2}\lambda(\varepsilon_1^e + \varepsilon_2^e + \varepsilon_3^e)^2 + \mu(\varepsilon_1^{e2} + \varepsilon_2^{e2} + \varepsilon_3^{e2}) + c^\circ \end{aligned} \quad \text{(A.23)}$$

and

$$\begin{aligned} \psi = \psi(\boldsymbol{\sigma}) &= \frac{1}{2E} I_1(\boldsymbol{\sigma})^2 + \frac{1+\nu}{E} I_2(\boldsymbol{\sigma}) + c^\circ \\ &= \psi(\sigma_1, \sigma_2, \sigma_3) = -\frac{\nu}{2E}(\sigma_1 + \sigma_2 + \sigma_3)^2 + \frac{1+\nu}{2E}(\sigma_1{}^2 + \sigma_2{}^2 + \sigma_3{}^2) + c^\circ. \end{aligned} \quad \text{(A.24)}$$

Equations $(A.23)_4$ and $(A.24)_4$ require some simple algebraic manipulation to be obtained. The same equations can otherwise be more easily obtained by writing eqs (A.17) and (A.18) in a coordinate system whose axes parallel the principal directions of $\boldsymbol{\varepsilon}^e$ and $\boldsymbol{\sigma}$ and by remembering that in such a system $\varepsilon_{ij}^e = \sigma_{ij} = 0$ for $i \neq j$, while $\varepsilon_{ii}^e = \varepsilon_i^e$ and $\sigma_{ii} = \sigma_i$. Of course, such a procedure exploits the well-known invariance property of eqs (A.17) and (A.18) with respect to the rotation of the coordinate Cartesian axes.

A.4 DISTORTIONAL AND VOLUMETRIC COMPONENTS OF ψ (ISOTROPIC CASE)

By taking the deviatoric and the spherical components of both sides of eq. (A.15), Hooke's law for isotropic elastic materials can also be expressed as:

$$s_{ij} = 2Ge^{e}_{ij}, \tag{A.25}$$

$$\bar{\sigma} = 3K\bar{\varepsilon}^{e}. \tag{A.26}$$

Here, $K = \lambda + 2\mu/3$ is the bulk modulus of the material. From these equations and from eqs (A.8), (A.10) and (A.11), we can express the deviatoric and the spherical components of ψ in terms of strain as follows (for some of the algebra involved cf. e.g. [2, pp. 473–479]):

$$\begin{aligned}\psi_{dev} &= \psi_{dev}(\boldsymbol{\varepsilon}^{e}) = G(\varepsilon^{e}_{ij} - \bar{\varepsilon}^{e}\delta_{ij})(\varepsilon^{e}_{ij} - \bar{\varepsilon}^{e}\delta_{ij}) + c^{\circ}_{dev} \\ &= \psi_{dev}(\varepsilon^{e}_{1}, \varepsilon^{e}_{2}, \varepsilon^{e}_{3}) = \frac{1}{3}G[(\varepsilon^{e}_{1} - \varepsilon^{e}_{2})^{2} + (\varepsilon^{e}_{2} - \varepsilon^{e}_{3})^{2} + (\varepsilon^{e}_{3} - \varepsilon^{e}_{1})^{2}] + c^{\circ}_{dev} \\ &= \psi_{dev}(\mathbf{e}^{e}) = Ge^{e}_{ij}e^{e}_{ij} + c^{\circ}_{dev} \\ &= \psi_{dev}(e^{e}_{1}, e^{e}_{2}, e^{e}_{3}) = G(e^{e2}_{1} + e^{e2}_{2} + e^{e2}_{3}) + c^{\circ}_{dev}\end{aligned} \tag{A.27}$$

and

$$\psi_{vol} = \psi_{vol}(\boldsymbol{\varepsilon}^{e}) = \psi_{vol}(\bar{\varepsilon}^{e}) = \frac{9}{2}K\bar{\varepsilon}^{e2} + c^{\circ}_{vol}. \tag{A.28}$$

In writing the last equation (A.27), the principal values of tensor $\mathbf{e}^{e}$ are denoted by e^{e}_{1}, e^{e}_{2} and e^{e}_{3}.

Alternatively, from eqs (A.4), (A.10), (A.11), (A.25) and (A.26) we can also express ψ_{dev} and ψ_{vol} in terms of stress as follows:

$$\begin{aligned}\psi_{dev} &= \psi_{dev}(\boldsymbol{\sigma}) = \frac{1}{4G}(\sigma_{ij} - \bar{\sigma}\delta_{ij})(\sigma_{ij} - \bar{\sigma}\delta_{ij}) + c^{\circ}_{dev} \\ &= \psi_{dev}(\sigma_{1}, \sigma_{2}, \sigma_{3}) = \frac{1}{12G}[(\sigma_{1} - \sigma_{2})^{2} + (\sigma_{2} - \sigma_{3})^{2} + (\sigma_{3} - \sigma_{1})^{2}] + c^{\circ}_{dev} \\ &= \psi_{dev}(\mathbf{s}) = \frac{1}{4G}s_{ij}s_{ij} + c^{\circ}_{dev} \\ &= \psi_{dev}(s_{1}, s_{2}, s_{3}) = \frac{1}{4G}(s^{2}_{1} + s^{2}_{2} + s^{2}_{3}) + c^{\circ}_{dev}\end{aligned} \tag{A.29}$$

and

$$\psi_{vol} = \psi_{vol}(\boldsymbol{\sigma}) = \psi_{vol}(\bar{\sigma}) = \frac{1}{2K}\bar{\sigma}^2 + c^{\circ}_{vol}. \tag{A.30}$$

Again, s_1, s_2 and s_3 are the principal values of **s**.

As already observed, it is usual to assume that the elastic energy of a virgin material – and hence its volumetric and deviatoric components – should vanish in the stress-free state. In applying the above formulae to virgin materials, one should in this case set $c^{\circ} = 0$, which also implies $c^{\circ}_{dev} = c^{\circ}_{vol} = 0$ since energy is a non-negative quantity.

From eqs (A.4), (A.6) and (A.27)–(A.30) it is not difficult to verify that:

$$\psi = \psi_{dev}(\mathbf{e}^e) + \psi_{vol}(\bar{\varepsilon}^e) \tag{A.31}$$

and

$$\psi = \psi_{dev}(\mathbf{s}) + \psi_{vol}(\bar{\sigma}) \tag{A.32}$$

Since the deviatoric and the volumetric components of a tensor are independent of each other, the above relations show that the deviatoric and the volumetric parts of ψ are actually uncoupled or orthogonal to each other. It should be stressed, however, that such a conclusion applies to isotropic elastic materials only. Uncoupled stress–strain relations (A.25) and (A.26) do not apply to non-isotropic elastic materials, because anisotropic elastic material will in general exhibit a cross coupling between the deviatoric and the volumetric components of stress and strain.

APPENDIX B

Rotation Tensors

Rotation tensors play an important role in this book due to the anisotropy of the subsequent yield surfaces. This prompts us to pay special attention to their definition and use. Sometimes, especially so when pursuing general arguments, the only property of rotation tensors that comes into play is that they are orthogonal tensors. However, in applying the results of this book, as well as in the derivation of some of its most crucial formulae, one has to be specific as to the way in which the components of a rotation tensor should be related to the rotation angles they mean to represent.

Rotation tensors are also involved in orthogonal coordinate transformations. Depending on the way we choose to define rotation tensors, the relevant coordinate transformation is performed by a certain operator or by its transpose. Failure to pay proper attention to this detail may spell confusion and lead to mistakes in the application of otherwise correct formulae.

For all these reasons, in this Appendix, we shall review the basic rules for the precise definition and consistent handling of rotation tensors. There are many textbooks on Classical Mechanics that cover the argument. The present review is based on the standard treatises [5] and [14], to which reference is made for a more complete treatment of the subject.

Readers should be advised that many different conventions can be adopted both in the definition of rotation tensors and in the choice of the independent parameters or angles they involve. Different books adopt different sets of conventions, which can sometimes make the comparison of the results they present difficult. The same references [5] and [14], quoted above, do not match entirely as far as these conventions are concerned. A comprehensive review of this topic is contained in reference [15]. It lists 24 different conventions for Euler's angles and the different bearing of each convention with the components of the rotation tensor. In such a situation, the most sensible suggestion to give the reader is also the obvious one: check the adopted convention and stick to it when interpreting or applying the relevant data. This Appendix explains the conventions adopted in this book as far as this matter is concerned.

B.1 THE TRANSFORMATION MATRIX

Let (x_1, x_2, x_3) and (x'_1, x'_2, x'_3) be two orthogonal coordinate systems sharing the same origin O in the ordinary three-dimensional Euclidean space. The triad of unit vectors that define the x_1-, x_2- and x_3-axes of the first system of coordinates will be referred to as **I**, **J** and **K**, respectively. Likewise, unit vectors **i**, **j** and **k** will denote the analogous triad of vectors relevant to the x'_1-, x'_2- and x'_3-axes of the second coordinate system. In what follows, the above two triads of vectors will also be referred to as **T** and **t**, respectively. More precisely, **T** and **t** will be understood to stand for the column matrices of orthonormal vectors defined by:

$$\mathbf{T} = \begin{bmatrix} \mathbf{T}_1 \\ \mathbf{T}_2 \\ \mathbf{T}_3 \end{bmatrix} \equiv \begin{bmatrix} \mathbf{I} \\ \mathbf{J} \\ \mathbf{K} \end{bmatrix} \quad \text{and} \quad \mathbf{t} = \begin{bmatrix} \mathbf{t}_1 \\ \mathbf{t}_2 \\ \mathbf{t}_3 \end{bmatrix} \equiv \begin{bmatrix} \mathbf{i} \\ \mathbf{j} \\ \mathbf{k} \end{bmatrix}, \tag{B.1}$$

respectively.

We shall take **T** as the fixed or initial triad, while **t** will be the rotated or the final one. The most fundamental way to specify the rotation of **t** with respect to **T** is to assign the direction cosines of vectors **i**, **j** and **k** with respect to triad **T**. This results in nine direction cosines that can be collected in a matrix **A** as follows:

$$\mathbf{A} = \begin{bmatrix} \mathbf{i}\cdot\mathbf{I} & \mathbf{i}\cdot\mathbf{J} & \mathbf{i}\cdot\mathbf{K} \\ \mathbf{j}\cdot\mathbf{I} & \mathbf{j}\cdot\mathbf{J} & \mathbf{j}\cdot\mathbf{K} \\ \mathbf{k}\cdot\mathbf{I} & \mathbf{k}\cdot\mathbf{J} & \mathbf{k}\cdot\mathbf{K} \end{bmatrix} \equiv \begin{bmatrix} A_{11} & A_{12} & A_{13} \\ A_{21} & A_{22} & A_{23} \\ A_{31} & A_{32} & A_{33} \end{bmatrix}. \tag{B.2}$$

Matrix **A** is also referred to as the *transformation matrix*. Its generic element A_{ij} is the cosine of the angle that the positive side of the x'_i-axis forms with the analogous side of the x_j-axis. Notice that the order of the indexes is not immaterial since, in general, $A_{ij} \neq A_{ji}$. It should also be observed that the transformation matrix could be equally well defined as the transpose of the above one. Both definitions appear in the literature and neither is standard. For instance, the transformation matrix defined in [5] is the transpose of that defined in [14]. Which definition is adopted may not matter when doing general analysis. A clear-cut distinction between them is however crucial when the transformation matrix is actually used to do the calculations.

From the orthonormality property of the considered triads it follows that:

$$\mathbf{A}\,\mathbf{A}^{\mathsf{T}} = \mathbf{1} = \mathbf{A}^{\mathsf{T}}\mathbf{A} \quad \text{or} \quad A_{ik}\,A_{jk} = \delta_{ij}. \tag{B.3}$$

It is well known that only six out of the nine equations on the far right-hand side (B.3) are independent. This means that the nine components of **A** can allbe defined in terms of just three independent variables; Euler's angles, to be introduced later, being a possible choice.

Equation (B.2) defines in an easy, complete and unambiguous way the rotation of the triad **t** with respect to the triad **T**. This is the rotation that has to be applied to **T** to bring it to coincide with **t**. The opposite rotation brings **t** to coincide with **T** and is represented by $\mathbf{A}^{\mathsf{T}}$. The proof can be obtained easily by observing that no rotation means $\mathbf{A} = \mathbf{1}$. But then, according to eq. (B.3), the subsequent application of **A** and its transpose produces no rotation. It must be concluded that the rotation brought about by $\mathbf{A}^{\mathsf{T}}$ is opposite to that produced by **A**.

It should also be observed that eq. (B.3) implies that $\det \mathbf{A} = \pm 1$, since $\det \mathbf{A} = \det \mathbf{A}^{\mathsf{T}}$ and $\det(\mathbf{A}\,\mathbf{A}^{\mathsf{T}}) = \det \mathbf{A} \det \mathbf{A}^{\mathsf{T}}$. Any proper rotation, however, should be such that $\det \mathbf{A} = 1$. This stems from simple continuity arguments once it is noted that any non-vanishing rotation must start from no rotation at all, that is from $\mathbf{A} = \mathbf{1}$. The determinant of the latter, however, is $\det \mathbf{1} = 1$. Since eq. (B.3) requires that $\det \mathbf{A}$ should be equal either to 1 or to -1, it follows that the relation $\det \mathbf{A} = 1$ must apply to *any* continuous rotation that starts from $\mathbf{A} = \mathbf{1}$. The case $\det \mathbf{A} = -1$ will apply when the rotation is accompanied by an odd number of axis inversions. In this case we speak of *improper rotation*.

As for the components of the rotated triad of vectors in the initial coordinate system (x_1, x_2, x_3), we can observe that from the definition of direction cosine it immediately follows that:

$$\begin{aligned} \mathbf{i} &= A_{11}\mathbf{I} + A_{12}\mathbf{J} + A_{13}\mathbf{K}, \\ \mathbf{j} &= A_{21}\mathbf{I} + A_{22}\mathbf{J} + A_{23}\mathbf{K}, \\ \mathbf{k} &= A_{31}\mathbf{I} + A_{32}\mathbf{J} + A_{33}\mathbf{K}. \end{aligned} \tag{B.4}$$

since **i**, **j** and **k** are unit vectors. In a more compact form, eq. (B.4) can be written as:

$$\mathbf{t}_i = A_{ij}\,\mathbf{T}_j; \tag{B.5}$$

which, in a still more compact form, reads:

$$\mathbf{t} = \mathbf{A}\,\mathbf{T}. \tag{B.6}$$

B.2 CALCULATING THE ROTATION OF THE PRINCIPAL TRIAD OF SYMMETRIC SECOND ORDER TENSORS

Let $\mathbf{n}^{(i)}$ be the unit vector that parallels principal direction i (i = 1, 2, 3) of any given symmetric second order tensor $\boldsymbol{\varepsilon}$. It is well known that the components $n_r^{(i)}$ of $\mathbf{n}^{(i)}$ are determined by the following matrix equation:

$$(\boldsymbol{\varepsilon} - \varepsilon_{(i)}\,\mathbf{I})\,\mathbf{n}^{(i)} = \mathbf{0}, \quad \text{(i not summed)}. \tag{B.7}$$

Here, **I** means unit tensor while the quantities $\varepsilon_{(1)} \equiv \varepsilon_1, \varepsilon_{(2)} \equiv \varepsilon_2, \varepsilon_{(3)} \equiv \varepsilon_3$ are the principal values of $\boldsymbol{\varepsilon}$, as determined from the following equation:

$$\det(\boldsymbol{\varepsilon} - \varepsilon_{(i)}\,\mathbf{I}) = 0, \tag{B.8}$$

(cf. e.g. [1, Sect. 1.2.3]). An immediate consequence of eq. (B.4) is that the rotation, say tensor **R**, to be applied to the reference axes of the coordinate system (x_1, x_2, x_3) to superimpose them onto the principal direction triad $(\mathbf{n}^{(1)}, \mathbf{n}^{(2)}, \mathbf{n}^{(3)})$ of $\boldsymbol{\varepsilon}$ is given by:

$$\mathbf{R} = \begin{bmatrix} n_1^{(1)} & n_2^{(1)} & n_3^{(1)} \\ n_1^{(2)} & n_2^{(2)} & n_3^{(2)} \\ n_1^{(3)} & n_2^{(3)} & n_3^{(3)} \end{bmatrix}. \tag{B.9}$$

B.3 ROTATING A VECTOR OR ROTATING THE COORDINATE AXES

Equations (B.4), (B.5) and (B.6) are three equivalent ways of expressing the unit vectors $\mathbf{t}_i$ of the rotated triad **t** in terms of the analogous vectors $\mathbf{T}_i$ of the initial triad **T**. Their main application is in the solution of the two classical problems recalled below.

(1) The first problem concerns rotating a vector and can be formulated as follows. By referring to coordinate system (x_1, x_2, x_3), fixed in space and defined by triad **T**, determine the vector $\mathbf{v}'$ obtained by making a given vector **v** rotate rigidly with a triad **t**, initially coinciding with the triad **T**.

To solve this problem, let

$$\mathbf{v} = v_1\mathbf{I} + v_2\mathbf{J} + v_3\mathbf{K} = v_i\mathbf{T}_i \tag{B.10}$$

be the vector to be rotated. After rotation, it will become the new vector $\mathbf{v}'$. In terms of the rotated triad, the latter will obviously be expressed by:

$$\mathbf{v}' = v_1\mathbf{i} + v_2\mathbf{j} + v_3\mathbf{k} = v_i\mathbf{t}_i, \tag{B.11}$$

because it is assumed that **v** rotates rigidly with **t** and that, initially, the latter coincided with **T**. By introducing eq. (B.5) into eq. (B.11), we get:

$$\mathbf{v}' = v_i A_{ij}\mathbf{T}_j. \tag{B.12}$$

This shows that in coordinate system (x_1, x_2, x_3) the components of $\mathbf{v}'$ are:

$$v'_j = v_i A_{ij}, \tag{B.13}$$

which answers the problem.

Equation (B.13) is often written in the form:

$$\mathbf{v}' = \mathbf{A}^{\mathsf{T}}\mathbf{v}, \tag{B.14}$$

where, according to the so-called direct notation of tensor analysis, the boldface letters stand for the array of components of the quantities they represent in the considered coordinate system. By referring to eq. (B.14), a simple application of the so-called 'Quotient Law' of tensor analysis suffices to prove that the linear operator **A** is in fact a second order tensor. This tensor is usually known as the *rotation tensor*. In the coordinate system (x_1, x_2, x_3), its components are given by the elements of matrix **A**, as defined by eq. (B.2). Since these components meet condition (B.3), tensor **A** is an orthogonal tensor.

The second classical problem that can be solved by means of eq. (B.6) concerns coordinate transformations. It can be expressed as follows.

(2) Determine how the components of any given vector **v** change as the coordinate system rotates from (x_1, x_2, x_3) to (x'_1, x'_2, x'_3).

The solution to this problem can be obtained easily by observing that eq. (B.6) can be solved by applying eqs (B.3) to give:

$$\mathbf{T} = \mathbf{A}^{\mathsf{T}}\mathbf{t}. \tag{B.15}$$

Or equivalently:

$$\mathbf{T}_{\mathrm{i}} = \mathrm{A}_{\mathrm{ji}}\mathbf{t}_{\mathrm{j}}. \tag{B.16}$$

By introducing this equation into eq. (B.10) we get:

$$\mathbf{v} = \mathrm{v}_{\mathrm{i}}\mathrm{A}_{\mathrm{ji}}\mathbf{t}_{\mathrm{j}}, \tag{B.17}$$

showing that the components of $\mathbf{v}$ in the coordinate system $(\mathrm{x}'_1, \mathrm{x}'_2, \mathrm{x}'_3)$ are given by:

$$(\mathrm{v}_{\mathrm{j}})' = \mathrm{A}_{\mathrm{ji}}\mathrm{v}_{\mathrm{i}}, \tag{B.18}$$

which solves the problem.

In tensor notation, eq. (B.18) is usually written as:

$$\mathbf{v}' = \mathbf{A}\mathbf{v}, \tag{B.19}$$

which is formally analogous to eq. (B.14), although $\mathbf{v}'$ and $\mathbf{v}$ now denote the components of the *same* vector in two different coordinate systems; system $(\mathrm{x}'_1, \mathrm{x}'_2, \mathrm{x}'_3)$ and system $(\mathrm{x}_1, \mathrm{x}_2, \mathrm{x}_3)$, respectively. In this context, a more consistent – though seldom adopted – notation would be $(\mathbf{v})'$ for $\mathbf{v}'$, as we did in writing eq. (B.18).

As they stand, eqs (B.14) and (B.19) might appear somehow embarrassing, since one of them has the transpose sign appended to **A** while the other has not. In fact they are both perfectly correct, since they represent two quite different operations. As the above discussion should make clear, the right formula to apply depends on whether we are using it in the context of problem (1) or in that of problem (2).

It should also be observed that in order to produce a relative rotation between a vector $\mathbf{v}$ and the triad $\mathbf{T}$, one can proceed in two ways. The first one is to apply rotation tensor $\mathbf{A}$ to vector $\mathbf{v}$, thus obtaining $\mathbf{v}'$ from eq. (B.14). The second way is to apply the opposite rotation $\mathbf{A}^{\mathsf{T}}$ to the triad $\mathbf{T}$, leaving $\mathbf{v}'$ fixed. In this case, eq. (B.19) would apply, but now the rotation of the coordinate system would be $\mathbf{A}^{\mathsf{T}}$ rather than $\mathbf{A}$. That is why we would get $\mathbf{v}' = \mathbf{A}^{\mathsf{T}}\mathbf{v}$, which is formally identical to eq. (B.14) though in fact obtained from (B.19).

B.4 EULER'S ANGLES

The rotation of a triad of vectors or coordinate axes defined by tensor **A** can be executed by means of three successive rotations about appropriate axes. The angles of these rotations are termed *Euler's angles*. Different sequences of rotations and different axes can be chosen to obtain the same final rotation defined by **A**. To each choice corresponds a different definition of Euler's angles. We shall adopt the following definition for Euler's angles ϕ, θ and ψ (see [5, Sect. 4.4] for further details):

ϕ: an anticlockwise rotation about the x_3-axis, transforming the initial triad of coordinate axes (x_1, x_2, x_3) into the new triad (ξ_1, ξ_2, x_3);
θ: a subsequent anticlockwise rotation about the newly obtained ξ_1-axis, transforming the triad (ξ_1, ξ_2, x_3) into a further triad (ξ_1, ξ_2', x_3');
ψ: a final anticlockwise rotation about the x_3'-axis transforming the above triad (ξ_1, ξ_2', x_3') into the final triad (x_1', x_2', x_3').

In reference [5, Appendix B], this convention is denoted as *x convention*. In the same reference, some further alternative conventions are examined and compared. The convention adopted in reference [14] is denoted as *y convention* in reference [5].

Euler's angles, as defined above, can be shown to be related to the transformation matrix **A** as follows (cf. [5, p. 147]):

$$\mathbf{A} = \begin{bmatrix} \cos\psi\cos\phi - \cos\theta\sin\phi\sin\psi & \cos\psi\sin\phi + \cos\theta\cos\phi\sin\psi & \sin\psi\sin\theta \\ -\sin\psi\cos\phi - \cos\theta\sin\phi\cos\psi & -\sin\psi\sin\phi + \cos\theta\cos\phi\cos\psi & \cos\psi\sin\theta \\ \sin\theta\sin\phi & -\sin\theta\cos\phi & \cos\theta \end{bmatrix}. \tag{B.20}$$

This serves to express **A** as a function of the three independent variables ϕ, θ and ψ.

B.5 EULER'S ANGLES CORRESPONDING TO A GIVEN ROTATION TENSOR

Equation (B.20) represents a set of nine scalar equations, which express each component A_{ij} of **A** in terms of Euler's angles as defined above. In the applications, we also need to solve the inverse problem: find the values of Euler's angles that correspond to a given rotation tensor **A**. In this case, the above set of eq. (B.20) can be treated as a system of nine equations in the three unknown variables ϕ, θ and ψ, the quantities A_{ij} being now assigned. Of course these equations are not independent of each other, as the orthonormality condition (B.3) entails six independent relationships among them. All in all, by extracting from eq. (B.20) the component equations corresponding to components $A_{13}, A_{23}, A_{33}, A_{31}$ and A_{32}, after some easy manipulations it is not difficult to conclude that:

$$\theta = \mathrm{atan}_2(\sqrt{(A_{13})^2 + (A_{23})^2}, A_{33}), \tag{B.21}$$

$$\phi = \mathrm{atan}_2\left(\frac{A_{31}}{\sin\theta}, -\frac{A_{32}}{\sin\theta}\right), \tag{B.22}$$

$$\psi = \mathrm{atan}_2\left(\frac{A_{13}}{\sin\theta}, \frac{A_{23}}{\sin\theta}\right). \tag{B.23}$$

Here, the two-variable function $\mathrm{atan}_2(y, x)$ is defined as the operator that returns the anticlockwise angle which the position vector relevant to point (x, y) forms with the positive x-axis. If variables x and y are real, then $\mathrm{atan}_2(y, x)$ lies in the interval $(-\pi, +\pi]$, thus justifying the term 'four-quadrant arc tangent' for this function. For instance, for $x = y = r$ we get $\mathrm{atan}_2(r, r) = \pi/4$, while for $x = y = -r$ we get $\mathrm{atan}_2(-r, -r) = 3\pi/4$. No such distinction between these two angles can be made if the more traditional single-variable function atan(x/y) is used instead.

Equations (B.21)–(B.23) enable us to determine Euler's angles corresponding to a given rotation tensor **A**.

B.6 CONTROLLING THE RELATIVE ROTATION OF TWO TRIADS OF AXES

A related problem is the following. Suppose we have two non-coaxial orthogonal triads of axes, say (x_1, x_2, x_3) and (x'_1, x'_2, x'_3), we have to determine the (proper) rotation tensor **D** that must be applied to the triad (x_1, x_2, x_3) in order to reduce the angles between the three pairs of corresponding axes (x_1, x'_1), (x_2, x'_2) and (x_3, x'_3) all by a factor $r \in [0, 1]$, arbitrarily assigned.

This problem has a unique solution if we assume, as we shall do, that the angles of the above pairs of corresponding axes lie in the interval $(-\pi/2, \pi/2]$. This case applies, in particular, whenever the above two triads represent the principal directions of any two symmetric second order tensors. As recalled in Sect. 4.2, a tensor of this kind is invariant for a π rad rotation about any of its principal axes. No restriction will be introduced, therefore, if we assume that the angle between any pair of corresponding principal axes of any two of such tensors is in the range $(-\pi/2, \pi/2]$.

Such an assumption has two important consequences. The first is that as the angle between the pair of corresponding axes is multiplied by a reduction factor $r \in [0, 1]$, the resulting angle will remain in the same quadrant as that of the original angle. This means, in particular, that the considered angle reduction process will not produce any change in the sign of the sine or the cosine of the reduced angle with respect to the sign the same functions attain for the original angle. The second consequence is that Euler's angles specifying the rotation of one triad of axes with respect to the other will be in the range $(-\pi/2, \pi/2]$ too.

In dealing with the problem at issue, we shall denote by $(\bar{x}_1, \bar{x}_2, \bar{x}_3)$ the final triad of axes, obtained by applying to the triad (x_1, x_2, x_3) tensor **D**, which is the main unknown of the problem. The rotation tensor needed to make the triad (x_1, x_2, x_3) superimpose onto the triad (x_1', x_2', x_3') will be denoted by **A**. Its components can be obtained from eq. (B.2), once the two triads (x_1, x_2, x_3) and (x_1', x_2', x_3') are assigned. Finally, the rotation that brings the triad $(\bar{x}_1, \bar{x}_2, \bar{x}_3)$ to superimpose onto the triad (x_1', x_2', x_3') will be denoted by $\bar{A}$. The adopted notation is illustrated in Fig. B.1.

To proceed further, let α_1, α_2 and α_3 be the angles between the pairs of conjugate axes (x_1, x_1'), (x_2, x_2') and (x_3, x_3'), respectively. These angles can either be assigned directly or be read off from transformation matrix **A**, if the latter is assigned instead of them. In this case, from eq. (B.2) we get:

$$\alpha_1 = \pm \operatorname{acos} A_{11}, \tag{B.24}$$

$$\alpha_2 = \pm \operatorname{acos} A_{22}, \tag{B.25}$$

$$\alpha_3 = \pm \operatorname{acos} A_{33}. \tag{B.26}$$

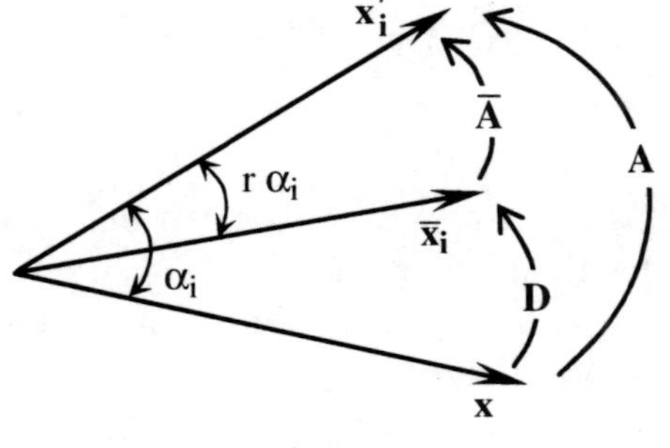

Fig. B.1 Adopted notation for angles and rotations of corresponding axes (the three axes referred to in the picture need not be co-planar).

If these equations are used, however, further information about the relative placement of the two triads is needed in order to eliminate the sign indeterminacy appearing on their right-hand sides. It should be remembered, however, that in our hypotheses all the above angles are in the range $(-\pi/2, \pi/2]$.

The solution of the present problem also requires that the initial values of Euler's angles giving the rotation of the triad (x'_1, x'_2, x'_3) with respect to the triad (x_1, x_2, x_3) should be known beforehand. These angles are immediately available from eqs (B.21)–(B.23), since rotation tensor $\mathbf{A}$ is known.

Our task is to determine the triad of axes $(\bar{x}_1, \bar{x}_2, \bar{x}_3)$, whose relative rotation with respect to the triad (x'_1, x'_2, x'_3) is such that the angles between the three pairs of axes $(\bar{x}_1, x'_1)$, $(\bar{x}_2, x'_2)$ and $(\bar{x}_3, x'_3)$ are given by:

$$\alpha'_1 = r\alpha_1, \quad \alpha'_2 = r\alpha_2, \quad \alpha'_3 = r\alpha_3, \tag{B.27}$$

respectively. This means that the diagonal elements of the transformation matrix $\bar{A}$ relating the triad $(\bar{x}_1, \bar{x}_2, \bar{x}_3)$ to the triad (x'_1, x'_2, x'_3) must be given by:

$$\bar{A}_{11} = \cos(r\alpha_1), \tag{B.28}$$

$$\bar{A}_{22} = \cos(r\alpha_2), \tag{B.29}$$

$$\bar{A}_{33} = \cos(r\alpha_3), \tag{B.30}$$

as immediately results from the structure of the transformation matrix, as specified by eq. (B.2). But, the above components are also related to Euler's angles as specified in eq. (B.20). It follows that:

$$\begin{cases} \cos\psi'\cos\phi' - \cos\theta'\sin\phi'\sin\psi' = \bar{A}_{11}, \\ -\sin\psi'\sin\phi' + \cos\theta'\cos\phi'\cos\psi' = \bar{A}_{22}, \\ \cos\theta' = \bar{A}_{33}, \end{cases} \tag{B.31}$$

where ϕ', θ' and ψ' are the three Euler's angles that define the rotation of the triad (x'_1, x'_2, x'_3) with respect to the triad $(\bar{x}_1, \bar{x}_2, \bar{x}_3)$.

In order to solve system (B.31) in terms of the unknown variables ϕ', θ' and ψ', we need to distinguish the case in which $\alpha_3 \neq 0$ from that in which $\alpha_3 = 0$. If $\alpha_3 \neq 0$, then $\bar{A}_{33} \neq 1$, as results from eq. (B.30). In this case, (B.31) is a system of three independent equations with the three unknown variables ϕ', θ' and ψ'. After some simple mathematics, they can be rearranged in the form:

$$\begin{cases} \cos(\phi' + \psi') = \dfrac{(\bar{A}_{11} + \bar{A}_{22})(1 - \bar{A}_{33})}{1 - (\bar{A}_{33})^2}, \\ \cos(\phi' - \psi') = \dfrac{(\bar{A}_{11} - \bar{A}_{22})(1 + \bar{A}_{33})}{1 - (\bar{A}_{33})^2}, \\ \qquad \cos\theta' = \bar{A}_{33} \neq 1. \end{cases} \tag{B.32}$$

From this system of equations, a unique solution can be obtained if we observe that in our hypotheses the sign of θ', $(\phi' + \psi')$ and $(\phi' - \psi')$ must be the same as those of the analogous unprimed quantities. From system (B.32) we get, therefore:

$$\begin{cases} \qquad \theta' = \operatorname{sgn}(\theta)\operatorname{acos}\left(\left|\bar{A}_{33}\right|\right), \\ (\phi' + \psi') = \operatorname{sgn}(\phi + \psi)\operatorname{acos}\left(\left|\dfrac{(\bar{A}_{11} + \bar{A}_{22})(1 - \bar{A}_{33})}{1 - (\bar{A}_{33})^2}\right|\right), \\ (\phi' - \psi') = \operatorname{sgn}(\phi - \psi)\operatorname{acos}\left(\left|\dfrac{(\bar{A}_{11} - \bar{A}_{22})(1 + \bar{A}_{33})}{1 - (\bar{A}_{33})^2}\right|\right), \end{cases} \tag{B.33}$$

where

$$\operatorname{sgn}(x) = \begin{cases} 1 & \text{if } x > 0, \\ 0 & \text{if } x = 0, \\ -1 & \text{if } x < 0. \end{cases} \tag{B.34}$$

We shall denote by l and m the right-hand sides of (B.33) and (B.33), that is:

$$l = \operatorname{sgn}(\phi + \psi)\operatorname{acos}\left(\left|\frac{(\bar{A}_{11} + \bar{A}_{22})(1 - \bar{A}_{33})}{1 - (\bar{A}_{33})^2}\right|\right) \tag{B.35}$$

and

$$m = \operatorname{sgn}(\phi - \psi)\operatorname{acos}\left(\left|\frac{(\bar{A}_{11} - \bar{A}_{22})(1 + \bar{A}_{33})}{1 - (\bar{A}_{33})^2}\right|\right). \tag{B.36}$$

From eq. (B.33) we therefore get:

$$\begin{cases} \theta' = \operatorname{sgn}(\theta)\operatorname{acos}\left(\left|\bar{A}_{33}\right|\right), \\ \phi' = \frac{1}{2}(l + m), \\ \psi' = \frac{1}{2}(l - m), \end{cases} \tag{B.37}$$

which solves system (B.31) for $\alpha_3 \neq 0$, or equivalently for $\bar{A}_{33} \neq 1$.

If $\alpha_3 = 0$, the rotation that makes the triad (x_1, x_2, x_3) superimpose onto the triad (x'_1, x'_2, x'_3) is simply a rotation about the x_3-axis. This means that in this case, $\alpha_1 = \alpha_2$. It also means that $\theta = \psi = \theta' = \psi' = 0$, as immediately results from the adopted definition of Euler's angles. Under these conditions, we have $\bar{A}_{11} = \bar{A}_{22}$ and $\bar{A}_{33} = 1$, as stems from eqs (B.28)–(B.30). Moreover, the first two equations (B.31) both reduce to the following equation:

$$\cos\phi' = \bar{A}_{11}. \tag{B.38}$$

As a consequence, Euler's angles relevant to the case $\alpha_3 = 0$ is given by:

$$\begin{cases} \phi' = \mathrm{sgn}(\phi)\mathrm{acos}\left(\left|\bar{A}_{11}\right|\right), \\ \theta' = 0, \\ \psi' = 0. \end{cases} \tag{B.39}$$

Equation (B.37) or eq. (B.39), as appropriate, enable us to calculate Euler's angles ϕ', θ' and ψ' in every case. By introducing them into eq. (B.20), the full expression of tensor $\bar{\mathbf{A}}$ that produces the rotation specified by these values of Euler's angles can be obtained immediately. From it, the desired rotation tensor $\mathbf{D}$ can finally be obtained from the following formula:

$$\mathbf{D} = \bar{\mathbf{A}}^{\mathsf{T}}\mathbf{A}, \tag{B.40}$$

as a glance at Fig. B.1 will justify at once.

The particular case in which the two triads only differ by a rotation about the x_3-axis is often of practical interest. In this case, $\alpha_3 = 0$ and eq. (B.39) applies. If we denote the rotation about the x_3-axis by θ_3, we have $\alpha_1 = \alpha_2 = \phi = \theta_3$. From eqs (B.28) and (B.39), it then follows that:

$$\begin{cases} \phi' = r\theta_3, \\ \theta' = 0, \\ \psi' = 0. \end{cases} \tag{B.41}$$

In such a case, a simple application of eq. (B.40) leads to:

$$\mathbf{D} = \begin{bmatrix} \cos\left[(1-r)\theta_3\right] & \sin\left[(1-r)\theta_3\right] & 0 \\ -\sin\left[(1-r)\theta_3\right] & \cos\left[(1-r)\theta_3\right] & 0 \\ 0 & 0 & 1 \end{bmatrix}. \tag{B.42}$$

APPENDIX C

Anisotropic Past Strain Effect

As an elastic-plastic material is plastically strained, its yield surface evolves differently depending not only on the extent of the past plastic strain but also on its direction. Dependence on the direction of the past plastic strain should not be confused with the anisotropy of the yield surface. The latter means that the elastic limit depends on the orientation of the material relative to stress. Past plastic strain direction dependence is a different phenomenon. It indicates that, as the yield limit is exceeded, the rule that controls the evolution of the yield surface – be this surface isotropic or not – will be different depending on the direction of the applied stress with respect to the direction of the plastic strain that the material suffered in the past. We shall refer to this phenomenon as the plastic evolution anisotropy relative to the past plastic strain, or the *anisotropic past strain effect* for short.

The best-known evidence of this phenomenon can be obtained from the uniaxial test. Take a specimen that was strongly prestrained in tension (compression) and then unloaded. If the specimen is reloaded in tension (compression), the σ/ε curve exhibits a sharp bend at the elastic-to-plastic transition. However, if the specimen is reloaded in compression (tension), the analogous elastic-plastic transition draws a sensibly more rounded bend in the σ/ε curve, no matter the extent of the previous plastic deformation in the opposite direction. A given uniaxial plastic strain increment, therefore, has a different effect on the evolution of the yield surface (in this case, produces a different sharpness in the elastic-plastic transition of the σ/ε curve and hence a different work-hardening effect) depending on the direction in which the material was prestrained. Such behaviour reveals that the plastic evolution of the material, i.e. its work-hardening rule, depends on the prestrain direction. This is an anisotropic past strain effect at work.

We have already discussed such a phenomenon in Sect. 7.7. The conclusion was that one should keep two separate accounts for the past plastic work in tension and in compression in order to model work-hardening behaviour under reverse loading. In the same section, it was noted that the phenomenon called for a more general work-hardening rule: a rule that, for a given deformation history, produces different evolutions of the yield surface depending on the orientation of the applied stress with respect to the previous plastic strain.

A simple work-hardening rule capable of embodying the anisotropic past strain effect is formulated below. Owing to the scarcity of experiments on this topic, we cannot even attempt to back it up with experimental evidence. The main purpose here is simply to provide an instance of how such an effect can be accommodated in the analysis. The proposed rule, moreover, should provide at least a primitive framework to direct and organize the much-needed experimental work on this otherwise hardly tractable subject.

To proceed further, we need to give a precise definition to what we loosely referred to above as orientation of stress (or, equivalently, of plastic strain increment) with respect to past plastic strain. To this end, we can greatly simplify the analysis by recalling that the spherical component of stress has no effect on plastic deformation and that, moreover, plastic strain is assumed to be deviatoric. We can accordingly limit our attention to the definition of the relative orientation (or angle) of two deviatoric second order tensors. A tensor of this kind is usually referred to for short as a *deviator*. We shall first of all assume that the three principal values of the considered deviators are not zero. Deviators possessing just one vanishing principal value will be dealt with later. (The case in which two principal values vanish is of little interest here, since it implies that the third principal value must vanish too, thus reducing the deviator to the null tensor.)

If all the principal values of a deviator are not zero, the magnitude of one of them must be greater than that of the other two. The sign of these two principal values, moreover, must be opposite to that of the principal value of greatest magnitude. All this follows immediately from the fact that a deviator is traceless. When no principal value vanishes, therefore, one principal value is predominant with respect to the other two and it is natural to define *director axis* (or, briefly, *director*) as the principal axis of the deviator that corresponds to the principal value of greatest magnitude. The angle between any two such deviators can then be defined as the angle between their director axes. In what follows, this angle will be denoted as χ.

The above definition does not specify which of the two angles formed by the two intersecting directors should be chosen as angle χ. In order to give an unambiguous definition of this angle, we shall stipulate that, if the considered directors correspond to principal values of the same sign, then χ is the angle between them that falls in the range $[-\pi/2, \pi/2]$; otherwise χ is the other angle, that is $\chi \in [\pi/2, 3\pi/2]$.

Once χ is thus defined, modelling the anisotropic past strain effect poses no serious problems. Let us refer, for example, to the evolution law expressed by eq. (7.2.3) and assume that eqs (7.3.1) and (7.3.2) apply. A way of introducing the anisotropic past strain effect into this law could be to redefine the variable $\tilde{w}$ appearing in eq. (7.3.3), so as to make it a function of angle χ between the actual stress at yield (or, more precisely, its deviatoric component $\mathbf{s}$) and the past plastic strain. The latter, however, is usually different at different times τ of the strain history, which spans from the initial time (that corresponds to $\tau = -\infty$) up to the present time (that is attained for $\tau = t$). For any given value of $\mathbf{s}$, therefore, different values of χ will be associated to different times of the past plastic strain history. A redefinition of the variable $\tilde{w}$, in which the effect of these different angles is taken into account, could be the following:

$$\tilde{w} = \tilde{w}(\mathbf{s}) = \int_{-\infty}^{t} g(\chi)\, \dot{w}^p \, d\tau. \tag{C.1}$$

As stated above, tensor $\mathbf{s}$ appearing in the equation is the deviatoric stress at yield, at the considered time t. As such, it must belong to the yield surface to which we intend to refer when applying eq. (C.1). In the same equation, $g(\chi)$ is an appropriate weighing function that is supposed to assume values in the interval [0,1], while $\dot{w}^p$ indicates the plastic power (the time rate of plastic work) per unit volume at time τ of the past plastic strain history. Angle χ appearing in eq. (C.1) is meant to be the angle between the considered value of $\mathbf{s}$ and the tensor $\dot{\boldsymbol{\varepsilon}}^p d\tau$, which represents the past plastic strain increment at time τ. It is this dependence of χ on $\mathbf{s}$ that makes $\tilde{w}$ depend on $\mathbf{s}$, as specified in eq. $(C.1)_1$. Clearly, the particular case in which $g(\chi)$ is constant and equal to 1, makes $\tilde{w}$ coincide with the past plastic work as usually defined, thus making eq. (C.1) coincide with the definition of $\tilde{w}$ we gave in Sect. 7.3.

Of course, the actual form of function $g(\chi)$ should be obtained from experiment. The following form, however, is a particularly simple one:

$$g(\chi) = \frac{1 + \cos\chi}{2}. \tag{C.2}$$

For want of a better choice, this expression for $g(\chi)$ appears to be acceptable, since it implies that $g(0) = 1$ and $g(\pi) = 0$. This means that a past plastic strain in a given direction is capable of affecting the evolution of the yield surface for yield stress in the same direction, but has no effect on the evolution of the yield surface for yield stress in the opposite direction. Such a feature is consistent with the experimental evidence on reverse loading in uniaxial tests, as already discussed in Sect. 7.7.

It should be clear, however, that there is no *a priori* reason to assume that the plastic evolution rule should embody eq. (C.2) or (C.1), or even be given by eq. (7.2.3). The right form of this rule can only be determined by experiment and can be different for different materials. The above formulation is only a simple example of how the anisotropic past strain effect can be modelled in practice.

According to the definition of a director, it is the principal value of the greatest magnitude that determines the director axis of a deviator. It does not matter whether or not this principal value is large in comparison with the other two principal values. As a consequence, the director bears no indication of how much a deviator is polarized in the direction of its director. Take the following two deviators, for instance:

$$\mathbf{d}' = \begin{bmatrix} d_1 & 0 & 0 \\ 0 & -\frac{1}{2}d_1 & 0 \\ 0 & 0 & -\frac{1}{2}d_1 \end{bmatrix} \quad \text{and} \quad \mathbf{d}'' = \begin{bmatrix} d_1 & 0 & 0 \\ 0 & -0.001\,d_1 & 0 \\ 0 & 0 & -0.999\,d_1 \end{bmatrix}. \tag{C.3}$$

Their directors are the same and coincide with the principal axis associated with the principal value d_1. Yet, if we compare the first and the third principal values of $\mathbf{d}''$, we find that their magnitude is almost the same. This indicates that $\mathbf{d}''$ is less polarized than $\mathbf{d}'$ along the director. Lamé's representation of these two tensors gives a clear picture of such a situation (Fig. C.1).

The following scalar D provides a suitable measure of *polarization* for any deviator $\mathbf{d}$:

$$D = \frac{|d_{max}| - |d_{min}|}{|d_{max}|}. \tag{C.4}$$

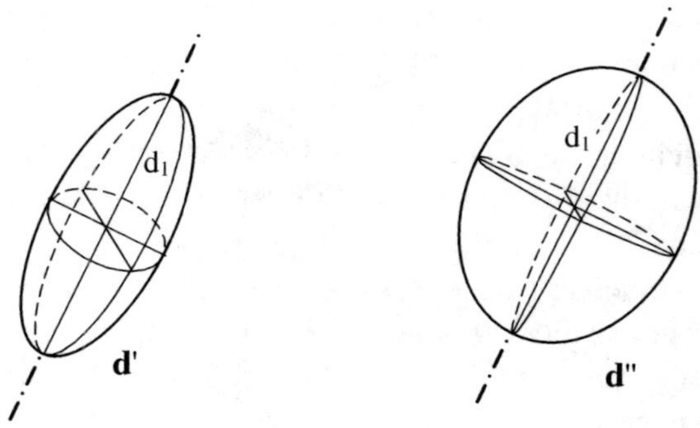

Fig. C.1 Lamé's representation of tensors $\mathbf{d}'$ and $\mathbf{d}''$ as defined by eq. (C.3). Ellipsoid $\mathbf{d}'$ is cigar shaped; exhibiting greater polarization along the same director (d_I) than tensor $\mathbf{d}''$. The latter is flat and dish-like.

Here, d_{max} and d_{min} are the greatest and the smallest principal values of $\mathbf{d}$, respectively. We shall refer to D as the polarization of $\mathbf{d}$ in the direction of its director. Since the trace of a deviator vanishes, it is an easy matter to prove that:

$$0 \leq D \leq 1 \tag{C.5}$$

for any deviator.

Once D is adopted as a polarization measure of a deviator in the direction of its director, it is only natural to expect a stronger anisotropic past strain effect from strongly polarized past strains. Such a feature can be embodied in the present approach by modifying eq. (C.1) as follows:

$$\tilde{w} = \tilde{w}(\mathbf{s}) = \int_{-\infty}^{t} g(\chi, D)\, \dot{w}^p \, d\tau, \tag{C.6}$$

where function $g(\chi, D)$ is supposed to be an increasing function of D, while the latter is meant to be the polarization of the past plastic strain $\dot{\boldsymbol{\varepsilon}}^p \, d\tau$ at time τ.

The above arguments refer to deviators possessing no vanishing principal values. The case of deviators with one vanishing principal value will now be considered. This is a practically important case, since it applies to pure shear stress and to simple shear strain. If one principal value of a deviator vanishes, the other two must be equal in magnitude and opposite in sign. This makes both the principal axes corresponding to these values entitled to be directors, according to the definition of director given above. A way out of this indeterminacy – though by no means the only way out – could be by observing that in this case, Lamé's ellipsoid degenerates to a circle. This means that (1) the polarization factor D of these deviators should vanish, and that (2) these deviators are plane tensors. In other words, such deviators have no polarization in their plane and, being plane tensors, have no component outside that plane.

These observations appear to justify assuming that such kinds of deviators are of no consequence as far as the anisotropic past strain effect is concerned. If this standpoint is adopted – which again should be confirmed by appropriate experiments – it would imply that function $g(\chi, D)$ appearing in eq. (C.6) should be such that $g(\chi, 0) = 0$, no matter the value of χ. This will make any deviator with vanishing polarization factor produce a vanishing contribution to $\tilde{w}$, which would also make it irrelevant as far as the anisotropic past strain effect is concerned.

It should also be observed that if $g(\chi, D)$ is a continuous function of D, then the assumption that $g(\chi, 0) = 0$ for every value of χ will make the value of g relevant to any deviator possessing one vanishing principal value join seamlessly, as D tends to zero, to the values of g relevant to the general case of deviators possessing $D \neq 0$.

Finally, the following consequence of the anisotropic past strain effect may be worth mentioning. This effect makes the yield surface of the material evolve differently depending on the direction of the plastic strain rate with respect to the previous plastic strain. If unrecognized, its presence may be a primary source of confusion when comparing the theoretically predicted yield surfaces with the experimentally determined ones. Especially so, if the experimental yield limit is based on a fixed strain offset (*strain offset method*). This can better be seen by remembering that the theoretical yield surfaces are defined with reference to the true elastic limit of the material, independent of any consideration about the amount of strain offset once this limit is trespassed. To determine the yield surface experimentally, on the contrary, we stipulate that the yield stress is the one that corresponds to a given – though usually small – plastic strain upon unloading (*offset strain*). If the evolution law of the yield surface is different in different directions, the same extent of offset will produce different errors in the evaluation of the elastic limit of the material, depending on the direction of the offset strain itself. The observed yield surface will consequently be distorted with respect to the true one, especially for strongly prestrained specimens, where the anisotropic past strain effect is usually large.

APPENDIX D

Recurring Rotation Tensors

It may help to recall here the meaning of some rotation tensors that are frequently used in this book. All of them define the rotation that has to be applied to the axes of the reference coordinate system in order to superimpose them onto a new triad of axes. The latter is different depending on the particular rotation tensor under consideration.

Since the reference coordinate system may itself be rotated, we shall distinguish between the *initial* coordinate system and the *rotated* coordinate system. The term *actual* coordinate system will indicate one or the other, depending on which of them is being assumed as the reference coordinate system.

The rotation tensors that most frequently recur in this book are the following:

- $\mathbf{Q}$ Rotation tensor defining the directions of the principal axes of any second order tensor with respect to the triad of axes of the actual coordinate system.
- $\mathbf{R}^\circ$ Rotation tensor defining the principal directions of $\boldsymbol{\varepsilon}^\circ$ with respect to the initial coordinate system.
- $\mathbf{R}$ Rotation tensor representing the rigid-body rotation of the material starting from its initial position (the one that is considered when defining $\mathbf{R}^\circ$).

P Rotation tensor defining the direction of the triad of axes of the rotated coordinate system with respect to the analogous triad of axes of the initial coordinate system.

S Rotation tensor defined as $\mathbf{Q}\,\mathbf{R}^{\circ\mathsf{T}}\,\mathbf{P}\,\mathbf{R}^{\mathsf{T}}$, cf. eq. (6.1.5). It represents the rotation that makes the triad of principal axes of $\boldsymbol{\varepsilon}^{\circ}$ superimpose onto the analogous triad of $\boldsymbol{\varepsilon}$. (Cf. also Sect. 4.7 for the particular case in which $\mathbf{P} = \mathbf{R} = \mathbf{1}$.)

References

[1] Lubliner, J., *Plasticity Theory*, Macmillan: New York, 1990.

[2] Paul, B., Macroscopic criteria for plastic flow and brittle fracture. *Fracture*, ed. H. Liebowitz, Academic Press: New York, Vol. 2, pp. 313–469, 1968.

[3] Leigh, D.C., *Nonlinear Continuum Mechanics*, McGraw-Hill: New York, 1968.

[4] Timoshenko, S.P., *History of Strength of Materials*, Dover, New York, 1983.

[5] Goldstein, H., *Classical Mechanics*, Addison-Wesley: Reading, MA, 1980.

[6] Fung, Y.C., *Foundations of Solid Mechanics*, Prentice-Hall: Englewood Cliffs, New Jersey, 1965.

[7] Paglietti, A., Eggs of any shape from a single virgin von Mises hen. *Progress in Mechanical Behaviour of Materials, Proc. 8th Int. Conf. Mech. Behaviour of Materials, Victoria, BC, Canada, 16–21 May 1999*, eds. F. Ellyin & J.W. Provan, Fleming Printing: Victoria, BC, Vol. 3, pp. 1089–1094, 1999.

[8] Paglietti, A., Von Mises virgin surface yields her secret while dancing in cold worked clothes. *Towards a History of Construction*, eds. A. Becchi, M. Corradi, F. Foce & O. Pedemonte, Birkhäuser: Basel, pp. 177–194, 2002.

[9] Phillips, A. & Lu, W.Y., An experimental investigation of yield surfaces of pure aluminum with stress-controlled and strain-controlled path of loading. *J. Engng. Mater. and Tech.*, Trans. ASME, **106**, pp. 349–354, 1984.

[10] Phillips, A. & Tang, J.L., The effect of loading path on the yield surface at elevated temperatures. *Int. J. Solids Struct.*, **8**, pp. 463–474, 1972.

[11] Ivey, H.J., Plastic stress-strain relations and yield surfaces for aluminum alloys. *J. Mech. Engng. Sci.*, **3**, pp. 15–31, 1961.

[12] Naghdi, P.M., Essenburg, F. & Koff, W., An experimental study of biaxial stress-strain relations in plasticity. *J. Appl. Mech.*, **25**, pp. 201–209, 1958.

[13] Sokolnikoff, I.S., *Mathematical Theory of Elasticity*, 2nd edn., McGraw-Hill: New York, 1956.

[14] Synge, J.L., Classical dynamics. *Encyclopedia of Physics*, Vol. 3, part 1, *Principles of Classical Mechanics and Field Theories*, ed. S. Flügge, pp. 1–225, Springer Verlag: Berlin, 1960.

[15] Craig, J.J., *Introduction to Robotics: Mechanics and Control*, 2nd edn., Addison–Wesley: Reading, MA, 1989.

[16] Paglietti, A., Thermodynamic nature and control of the elastic limit in solids. *Int. J. Non-Linear Mech.*, **24**, pp. 571–583, 1989.

Index